BIOLOGY MADE SIMPLE

Rita Mary King, Ph.D.
Assistant Professor of Biology
The College of New Jersey
with
Frances Chamberlain

Edited by
Q. L. Pearce
with
William J. Pearce, Ph.D., Professor of Physiology,
Pharmacology, and Biochemistry
Loma Linda University School of Medicine
Loma Linda, California

Illustrated by Scott Nurkin

A Made Simple Book
Broadway Books
New York

Produced by The Philip Lief Group, Inc.

Printed in the United States of America

Produced by The Philip Lief Group, Inc.
Managing Editors: Judy Linden, Lynne Kirk, Hope Gatto.
Design: Annie Jeon.

Visit our website at www.broadwaybooks.com

Library of Congress Cataloging-in-Publication Data

King, Rita M.
Biology made simple / Rita King.—1st Broadway Books ed.
p. cm.

1. Biology—Popular works. 2. Human biology—Popular works. I. Title.

QH309.K54 2003
570—dc21 2003041908

ISBN 0-7679-1542-9

10 9 8 7 6 5 4 3 2 1

CONTENTS

INTRODUCTION: UNIFYING THEMES OF BIOLOGY

KEY TERMS

biology	organism	environment
atom	element	cell

One of the most remarkable things about our home planet is that it is teeming with life. No matter where you travel from equator to pole, from mountain to ocean, you will encounter living things. This book introduces you to **biology**, the study of living organisms and their characteristics.

How is life defined? This is a question open to some debate, but biologists have determined certain traits that organisms have in common. Life is associated with various properties and processes: order, growth and development, reproduction, energy utilization, response to the environment, homeostasis, and evolutionary adaptation. From the smallest bacterium to the majestic blue whale, living things exhibit these properties and processes in some form.

TRAITS OF LIVING THINGS

Order: The basic cell is composed of smaller units. Living things are made up of cells. An organism may be made of a single cell, or a complex combination of cells.

Growth and Development: Living organisms increase in size and/or number of cells.

Reproduction: Through sexual or asexual means, living things reproduce new organisms of their species and so pass on hereditary material.

Energy Utilization: Through a process called metabolism, living things utilize an energy source (food) to fuel functions.

Response to Environment: Living things respond to stimuli in the environment. An example of a response is movement toward food or away from a threat.

Homeostasis: Living things seek an internal environment that is favorable to cell function.

Evolutionary Adaptation: Living things may adapt to environmental change resulting in an increased ability to reproduce.

As amazing as it seems, you share a great deal with the roses in your garden, your neighbor's goldfish, and the "bug" that gave you the sniffles a few weeks ago. Even though organisms may seem very different on the outside, there are many unifying themes in biology. We will explore them further in later chapters, but let us begin our journey with an introduction to these themes.

HIERARCHICAL ORDER IN LIFE

The Greek philosopher, Democritus of Abdera (460–370 BC), proposed that all forms of matter are made up of basic, indivisible particles, which he called atoms. He was off to a good start, but it would take more than two thousand years for scientists to come up with a working model.

Democritus might have been surprised to learn that atoms are made up of even smaller units. To understand the order of living things, let's begin at the subatomic level. The principal subatomic particles are protons, neutrons, and electrons. Atoms make up all known substances. Some substances, called elements, consist entirely of one type of atom (See Appendix—Periodic Table of Elements). Elements may combine to form molecules. When different elements combine, they produce chemical compounds. We'll take a closer look at elements, molecules and compounds in chapter two.

Biologically speaking, certain atoms combine to form complex biological molecules, such as proteins. Various biological molecules form cellular organelles, such as a nucleus, and these organelles are ordered into cells. The cell is the smallest unit of life, but it is just the beginning. In multicellular organisms, similar cells that perform a specific function are organized into tissues. Different tissues that work together to perform a particular function become organs, the stomach for example. An organ system is a group of organs that works together to perform a particular function, such as digestion. The result is an individual complex organism, such as a human.

Let's not stop there. Levels of organization in biology go far beyond the individual organism. A localized group of organisms belonging to the same species is called a population, and populations are grouped into a community. An ecosystem is an energy-processing system of community interactions that include non-living (abiotic) environmental factors such as, water, air, and soil. Biomes are large-scale communities classified by a main vegetation type and distinctive combinations of plants and animals. Finally, the biomes make up the biosphere—that part of the planet on which all life exists.

LEVELS OF ORGANIZATION

Biosphere: Regions of Earth's crust, water, and atmosphere inhabited by living things

Ecosystem: A community and its environment

Community: Populations of different species living in the same area

Population: Group of organisms of the same species living in the same area

Multicelled Organism: An individual organism made up of cells organized as tissues, organs, and organ systems

Organ System: Two or more organs interacting to perform a particular function

Organ: A unit composed of tissues, combined to perform a specialized task or tasks

Tissue: Cells and substances combined to perform a specialized function

Cell: Smallest unit with the ability to live and reproduce

Organelle: Intracellular compartment surrounded by a protective membrane

Molecule: A unit made up of two or more atoms of the same or different elements

Atom: Smallest unit of an element that retains the properties of the element

Subatomic Particle: Fundamental unit of matter such as proton, neutron, and electron

CELLULAR BASIS OF LIVING THINGS

In the Hall of Fame for Biological Theories, the Cell Theory is among the super stars. Its seeds were planted during the seventeenth century following the invention of the microscope. Robert Hooke (1635–1703) was well known in his day as the Curator of Experiments for the Royal Society of London. He had developed an excellent compound microscope and he used it to observe small organisms. Hooke noted box-like structures in a thin slice of cork. He called them cells since they reminded him of the cells in a monastery. Within a decade, the Royal Society asked Hooke to review the work of Antonie van Leeuwenhoek. The latter had observed living cells when he viewed microbes in pond water.

In 1665, Robert Hooke published a book of his observations called, *Micrographia*. He illustrated it with detailed drawings of microscopic views of insects, plants, feathers, and, yes, those famous cork cells. Samuel Pepys praised it as, "... the most ingenious book that I ever read in my life."

In 1838, Matthias Schleiden proposed that, like cork, all plants were composed of cells. A year later, Theodor Schwann took it even further. He believed the cell was the basic functioning unit of all living things. The cell theory was officially born in 1868, when pathologist Rudolf Virchow combined the earlier ideas and went on to show that cells are the result of division of preexisting cells.

"Life will always remain something apart, even if we should find out that it is mechanically aroused and propagated down to the minutest detail."

Rudolf Virchow, 1855

The cell is the lowest level of biological structure capable of performing all of the activities of life. The invention of the electron microscope in the 1950's has allowed us to determine the ultrastructure of cells. All have a plasma membrane and, at some stage in their development, contain deoxyribonucleic acid, better known as DNA. Structurally, cells can be divided into two groups, prokaryotes and eukaryotes.

Prokaryotes were the first cells on Earth. Archaebacteria and bacteria are prokaryotic cells. They are generally between 0.1 and 10 μm in size. Their DNA is not separated from the rest of the cell by a membrane-bound nucleus and they lack membrane-bound organelles. Most prokaryotic cells have tough external cell walls. There is no cytoskeleton (a lattice-like network of proteins that holds a cell together and gives it its shape) or cytoplasmic streaming (the flow of cytoplasm from one cell region to another). Metabolism can take place in the presence of oxygen (aerobic) or not (anaerobic), depending upon the cell type and species. Cell division is accomplished by binary fission.

- **μm = micrometer or one millionth of a meter.**

The eukaryotic cell evolved about 1.5 billion years ago. Protists, fungi, plants, and animals have eukaryotic cells. These cells are generally larger than 10 μm. Most of the genetic material is found within a membrane-bound nucleus, and membrane-bound organelles are present. A cell wall occurs in some of the species, but its chemical structure differs dramatically from that of prokaryotic cells. A cytoskeleton is present and there is cytoplasmic streaming. Metabolism is generally aerobic. Cell division is either by mitosis or meiosis (see Chapter 5).

There are similarities between prokaryotic and eukaryotic cells, too. All cells are bounded by a plasma membrane that encloses proteins and usually nucleic acids such as DNA and RNA. Together, these components guide the cell's energy sources through metabolic activities controlled by specialized proteins called enzymes.

- **Cytoplasm is the gel-like fluid within a cell.**

DNA-HERITABLE INFORMATION

DNA is a familiar term in the twenty-first century. It is a focal point of research, a key element in many court cases, and even a featured player in television mysteries. The basic chemical components of DNA are the same for all organisms. It is a double helix, or twisted ladder, made up of sugar and phosphate molecules with four nitrogenous bases: adenine, guanine, cytosine, and thymine. Sugar-phosphate forms the twisted edge and the bases are the rungs of the ladder. The linear sequence of these bases encodes information in a gene—a specific length of DNA. Inheritance is based on a complex mechanism for copying DNA and then passing the DNA from parent to offspring. All forms of life use essentially the same genetic code (See Appendix—The Genetic Code). Each nucleotide sequence codes for the synthesis of one particular protein, regardless of the cell or species in which it occurs. Differences in organisms are due to differences in their nucleotide sequences.

- **A nucleotide is made up of a nitrogenous base, a five-carbon sugar, and a phosphate group. On our twisted ladder, each nucleotide represents half a rung.**

STRUCTURE AND FUNCTION

There is a relationship between structure and function at all levels of biological organization. Typical functions in multicellular organisms include structural support, sensory processing, and movement. All of these specialized functions require unique structures optimized to serve these activities. The small intestine of humans is an excellent example. It is the major site of digestion and absorption of nutrients, and is designed to provide maximum surface area for this to take place. The six-meter long small intestine is folded into loops. The inner surface is lined with microscopic projections called villi, which in turn are covered with tiny appendages called microvilli.

Another example of structure/function is the anatomy of plant leaves, specifically those plants that inhabit dry, arid regions. Most succulent plants have leaves modified for storing water.

INTERACTIONS OF ORGANISMS

Have you ever had the urge to get away and be all by yourself? Technically, that's an impossibility. You are always in contact with living organisms of one kind or another. Organisms usually exist in open systems and interact constantly with each other and with their environments. An example is the food web. Plants, as well as certain bacteria and protists, manufacture most of the food consumed by other organisms and thus make up the base of most food webs. These food webs represent the flow of energy that begins with the sun, runs through plants and other photosynthetic organisms, and ultimately fuels complex multicellular predatory animals. These food webs, in turn, are often organized into ecosystems that can involve millions of organisms across thousands of species.

UNITY IN DIVERSITY

The commonality of life on Earth is demonstrated by a universal genetic code, similarities of cell structure, and identical metabolic pathways in many different species. Despite this commonality, there is great diversity in life, as represented by the 1.5 million species that have been named and identified so far. How many species there are remains unknown and could be many millions given the tremendous variety of organisms and their constant tendency to change. Scientists classify these organisms using the binomial, or "two name" system of nomenclature developed by Carolus Linnaeus in the eighteenth century. From most inclusive to most specific, scientists classify organisms into a kingdom, phylum, class, order, family, genus and species.

EVOLUTION—THE CORE THEME OF BIOLOGY

Evolution is a core-unifying theme of biology. Life evolves and species change over time in direct response to changes in environment. In 1859, Charles Darwin wrote *On the Origin of Species by Means of Natural Selection*. This very controversial book sold out the very first day of publication! Darwin suggested natural selection as a mechanism of evolutionary change. According to the twentieth century geneticist, Theodosius Dobzhansky, nothing in biology makes sense without evolution. It is generally believed that all life on Earth has descended from prokaryotic cells. We will delve further into the details of Darwin's elegant theory in chapter eight.

SCIENTIFIC PROCESS

As long as people have wondered why things are the way they are, they've invented ideas to explain the world around them. Many of these ideas have proven to be far-fetched, but everything we know today is the direct result of someone's curiosity and imagination. So how do we sort out the difference between the far-fetched ideas and the truth? Simply put, we use the scientific method, the application of common sense and logic.

Science depends on facts gathered from observation. Using the scientific process, an investigator asks a question and forms a tentative answer, or hypothesis, then tests the prediction by doing experiments and making new observations. Then the cycle begins anew: new observations, fresh hypotheses, more experiments. Through this cycle, called the scientific method, researchers are slowly but most certainly demystifying the many puzzles that underlie who we are and how life works.

THE SCIENTIFIC METHOD

OBSERVATION: The investigator makes observations, studies previous data, and defines the problem.

HYPOTHESIS: The investigator drafts one or more testable statements.

EXPERIMENT: The investigator designs and conducts controlled experiments and makes further observations.

CONCLUSION: The investigator analyzes the results. The hypothesis is supported or rejected.

Scientists, however, don't go looking for absolute proof. Hypotheses can be eliminated, but not confirmed with absolute certainty. Let's take a look at an example of the scientific method in action.

Observations	Tomato plants dying, leaves mottled, and contagion typical of viral disease.
Hypothesis	Virus X is killing the tomato plants.
Experiments	a.) Use electron microscope to verify presence of Virus X. b.) Infect healthy tomato plants with Virus X to see if mottled leaves occur and death occurs to the plants.
Results	Will support or refute hypothesis

- **Experiments always need a control group for comparison to the experimental group.**

SUMMARY

- Life is organized at different levels—atom, cell, organism, and above the organism.
- At every level in life's hierarchy, the whole is greater than the sum of its parts.
- The parts have precise structures and these dictate their functions.
- The double helix of DNA is the unifying chemical of life; its linear sequence defines the diversity of living things.
- Evolution, the modification of species, is the core theme of biology.
- The scientific method is a process whereby natural phenomena are believed to have natural causes. Biologists follow this process in their continuing quest to learn more about living organisms.

CHAPTER 2 CHEMISTRY FOR THE BIOLOGIST

KEY TERMS

matter	compounds	chemical bonds
organic	carbohydrate	protein
lipid	nucleic acid	

Imagine you are toasting marshmallows over a campfire on a late summer evening. Fireflies twinkle nearby. Do you find yourself pondering the chemical reactions that power the firefly's lamp, turn firewood to ash, and marshmallows to goo? Don't be concerned. It isn't that unusual. Look into the heart of a biologist and you will find a soft spot for chemistry. Living things are made up of a cornucopia of organic chemicals, and chemical reactions are ever-present in the natural world, from digestion to reproduction. To understand these processes we must journey to the world of the atom.

ATOMIC STRUCTURE

Matter is anything that takes up space and has mass. Mass is the amount of matter a substance has. Let's say you have a balloon and a bowling ball of the same size. If you were to drop them both on your foot, it would be painfully clear which has more mass. This brings up another point. Weight is a measure of gravitational force pulling on an object. The more mass something has, the more it weighs.

All matter, whether solid, liquid, or gas is made up of atoms, and all atoms have the same basic structure. Let's take helium for example (See Figure 2.1).

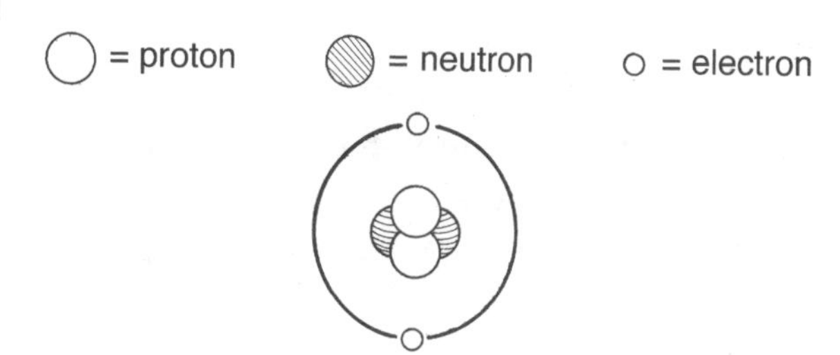

Figure 2.1—Diagram of a Helium Atom

The atom is composed of a nucleus at its center, which contains positively charged particles called protons. If neutrons (neutral—having no charge) are present, they will also be in the nucleus. They usually match up with protons in about a one to one ratio. Helium has two protons and two neutrons. Protons and neutrons each have an atomic mass of one dalton.

> The unit of measurement called a dalton is named after John Dalton (1766–1844). The English chemist measured the mass of certain atoms. He determined that each element was made up of its own kind of atom, and atoms of different elements had different properties and masses. He published many of his ideas in the *New System of Chemical Philosophy* in 1803.

Orbiting around the nucleus at very high speeds are tiny, negatively charged particles called electrons. They don't travel in a set path as a planet orbits a sun. Electrons occupy a space called a shell, or energy level, around the nucleus. There are multiple shells increasing in radius from the inner to outer levels. A shell can hold a limited number of electrons. For example, the innermost shell can hold only two. The next can hold eight electrons, and the next eighteen, and so on. Each shell consists of one or more orbitals, regions where one or two electrons might be found.

Atoms have an equal number of positively charged protons and negatively charged electrons and are neutral. If an atom gains or loses an electron, it becomes a charged atom called an ion. Ions have charges corresponding to the number of electrons lost or gained.

SUBATOMIC PARTICLES

Name	Symbol	Charge
neutron	n	0
proton	p	+1
electron	e	−1

ELEMENTS AND COMPOUNDS

An element is a substance made up of one kind of atom. It can't be broken down into a simpler substance. At present, scientists have identified 113 elements. (See Appendix—Periodic Table of Elements.) More await discovery. Admittedly, those with an atomic number greater than ninety-two are rarely found outside of a laboratory.

The Periodic Table of the Elements is an elegant example of compact information display. It was developed by Russian scientist, Dimitri Mendeleev, in 1869. To read the table, choose an element, let's use helium again, and start with the number at the top of the square. That is its atomic number and it denotes the number of protons in the nucleus. Each element is represented by a chemical symbol—a shorthand name. Take another look at helium. The number below the symbol is 4.00. That is its atomic mass—average mass of an element in all of its forms. The atomic mass is the sum of the masses of the protons and neutrons.

In nature, many elements have more than one atomic form due to the gain or loss of neutrons. These are called isotopes. Their chemical properties remain generally the same but their atomic mass changes.

An isotope is radioactive if the nucleus decays spontaneously giving off particles and energy. If the decay leads to a change in the number of protons, the atom of another element is formed. Radioactive isotopes have been used in scientific research for decades, and as diagnostic tools in medicine.

- **The mass of an electron is so small, it is not considered in the mass of the atom.**

Most elements are solids and most of them are metals. The elements carbon, oxygen, hydrogen, and nitrogen make up ninety-six percent of the chemical composition of living matter. When elements combine to form substances consisting of two or more different elements in definite proportions, chemical compounds are produced. A compound has different physical and chemical properties than its parent elements.

CHEMICAL BONDS

The old saying that opposites attract is certainly true at the atomic level. Charged particles are always ready for a little bonding. Elements combine to form substances of two or more different elements (compounds) through chemical bonds.

Chemical properties of an atom depend mostly on the number of electrons in its outermost shell—the valence electrons. Atoms are stable when their outer shell is complete. To achieve this, an atom will share, gain, or lose electrons. Sharing at least one pair of electrons is known as covalent bonding. The atoms involved end up with complete outermost valence shells, so it is a strong relationship. The bonds can be single, double, or triple depending upon how many electron pairs are being shared. The more pairs the atoms share, the more stable they become.

> Atoms with a filled outer shell are inert, or unreactive. You'll find them on the right edge of the Periodic Table. They are also called rare or noble gases.

Covalent compounds are usually liquids or gases at room temperature. In the elemental state, hydrogen, nitrogen, oxygen, and the halogens (the row to the left of the noble gases) are usually paired, or diatomic, due to covalent bonding. Covalent bonds are found in carbon dioxide, water, methane, and glucose, all compounds important to living things.

- **Covalent bonds are most likely to form among nonmetals that are similarly attractive to electrons.**

An electronegative atom attracts electrons of a covalent bond toward its own nucleus. This can make one atom in a bonded pair a little more "selfish" than the other. It tends to hang on to the shared electron more than its partner does. This gives the atom a slight negative charge, or polarity. When electrons are not shared equally, the result is a polar covalent bond (See Figure 2.2). Oxygen is one of the most electronegative elements known. As such, it often forms polar covalent bonds. This type of bonding is seen in a molecule of water.

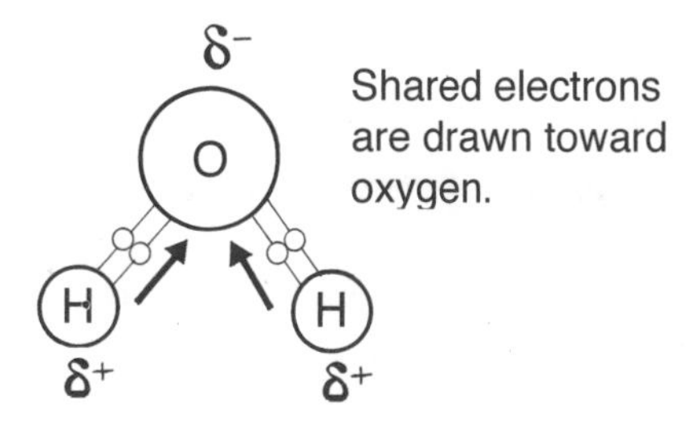

Hydrogen bonds form between water molecules

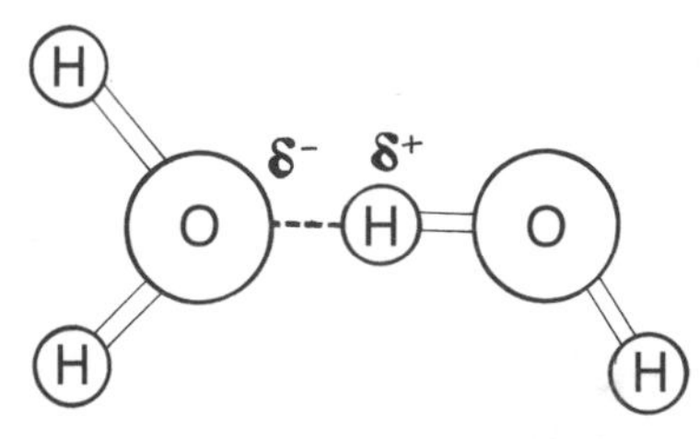

Figure 2.2—Unique Structure of Water

> Van der Waals interactions, named for physicist J.D. van der Waals, are weak attractions between molecules or parts of molecules that are brought about by localized charge fluctuations. These forces are important in the structure of proteins. Even weak bonds can have an important effect. Weak hydrogen bonds hold the nucleotide bases of DNA together.

Salt is an excellent example of an ionic bond—a bond in which electrons transfer from one atom to another. Sodium has a single electron in its outer shell. Chloride is one electron short in its outer shell. When an electron from sodium is transferred to chloride, they each achieve the sought after full outer shell. The electron shedding sodium atom becomes a positively charged ion and the electron grabbing chloride atom becomes a negatively charged ion. Remember, opposites attract! The two atoms join in an ionic bond that results in a compound known as sodium chloride (common table salt).

- **Ionic bonds are easily broken in water.**

CHEMISTRY OF WATER

As a molecule, water has unique chemical and physical properties that contribute to the survival of life on Earth. Electrons are not shared equally between the hydrogen and oxygen atoms of water, but are unevenly distributed (see Figure 2.2). Therefore, water is slightly charged at each end. The hydrogen atoms in a water molecule are slightly positive and thus are attracted to the slightly negative oxygen atoms in other water molecules. These interactions form strong, non-covalent hydrogen bonds that order the molecules to a higher level of structural organization.

- **Within ice, the molecules are highly organized in a lattice formation. Because of this, ice is less dense than water.**

Properties of Water

The next time you sip a glass of water consider this. Water boasts five physical properties that make it a molecule unlike any other: versatility as a solvent, cohesive behavior, imbibition, ability to stabilize temperature, and expansion upon freezing.

Water is the solvent of life. Ionic compounds dissolve in water. Generally, polar compounds are water-soluble. Hydrogen bonding gives liquid water more structure than most liquids. Therefore, the water molecules are more bound to each other.

- **Hydrophobic substances (e.g., nonpolar compounds) do not dissolve in water. Hydrophilic substances do dissolve in water.**

Cohesion is the force of attraction between like molecules. Have you ever overfilled a glass of water just a little so that the surface bulges but doesn't spill? That's the result of surface tension. Hydrogen bonds at the surface of the water form a sort of "skin" and some molecules are hydrogen-bonded to water molecules below the surface. Hydrogen bonds are the reason you can make a stone skip across a lake's surface! This cohesive behavior helps in the transport of water in a plant's xylem. Adhesion is the attraction between molecules of different materials, and it enables water to cling to the xylem walls and counter the effects of gravity.

> If you enjoy gardening, you have benefited from imbibition. It is the uptake of water by materials that don't dissolve in water. The embryo of a seed plant absorbs water, swells and bursts through the seed coat, allowing the seed to germinate.

Specific heat is the amount of heat that must be absorbed or lost for one gram of a substance to change its temperature by 1°C. Water has a high specific heat (1cal/g/°C), so it resists temperature changes. Hydrogen bonds release heat when they are formed and absorb heat to break. Therefore, water is able to stabilize temperature. This becomes very important when we realize that seventy-five percent of our planet is covered by water. This property of water moderates the Earth's climate, and modifies the transition between seasons in areas near large bodies of water. It stabilizes temperature in aquatic systems, and helps organisms release heat by evaporative cooling (sweating).

You probably learned in elementary school that substances expand when they are heated and contract when cooled. Fortunately for us, water doesn't do that. Due to the hydrogen bonding between water molecules, it does just the opposite. The result is that ice is less dense than water. It floats on top of the surface, releasing heat to the water below and acting as an insulator. And what would happen if it didn't do that? For one thing, bodies of water would freeze from the bottom up, making life pretty tough for anything living in them!

ACIDS, BASES, AND pH

Acids and bases are chemical opposites. An acid is a substance that forms or releases hydrogen ions when it dissolves in water. A base is a substance that reduces the relative hydrogen ion concentration, forming an alkaline solution. Strong acids (hydrochloric acid) and bases (sodium hydroxide) dissociate in water completely. Weak acids (carbonic acid) and bases (ammonia) dissociate only partially and reversibly.

A buffer is a substance that minimizes large sudden changes in pH. Buffers work by accepting hydrogen ions from solutions when they are in excess, and by donating hydrogen ions to the solution when they have been depleted.

The pH scale (See Figure 2.3) is used to assess the degree of acidity or basicity (alkalinity) in a solution using a scale ranging from 0 to 14. The letters pH stand for "potential of hydrogen." A pH of 7.0 is considered neutral. Below pH 7 is acidic and above pH 7 is basic, so a strong acid has a low pH. Most biological fluids have a pH range between 6.0 and 8.0. A notable exception is the acidic pH of your stomach, which has an important role in the digestion of proteins.

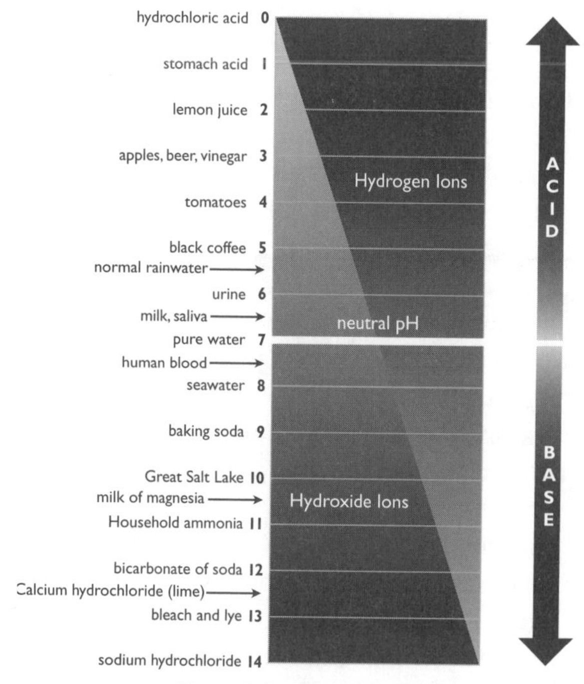

Figure 2.3—The pH Scale
This table shows the proportion of hydrogen ions to hydroxide ions at various pH levels. Levels between 0 and 7 are called *acidic;* levels between 7 and 14 are called *basic.*

- **The pH scale is logarithmic. Each measure represents a tenfold change. For example, something that is pH 2 is ten times more acidic than something that is pH 3.**

IMPORTANCE OF CARBON

The term organic describes any carbon compound (other than carbon oxides and carbonates). Carbon is the main element of organic compounds, which make up the organisms on this planet.

Carbon usually forms four covalent bonds in compounds due to its four valence electrons. Carbon skeletons may vary in length, shape (straight chain, branched, or ring), number and location of double bonds, and elements bonded to available sites. Frequent elements bonded to carbon include hydrogen, oxygen, and nitrogen.

- **Inorganic chemistry is the study of elements and compounds other than most carbon compounds. Although life as we know it would not exist without carbon, the study of the element itself falls within the realm of inorganic chemistry.**

In organic chemistry, molecules that have similar properties are classified as functional groups. When functional groups are bonded to the carbon atom, the resulting compounds wind up with specific chemical and physical properties.

- **Aldehydes** (having C=O at the end of a carbon chain) and ketones (having C=O in a place other than the end of a carbon chain) are important in sugars.
- **The amino group** (-NH_2) is an important functional group in amino acids, and hence, proteins. It is polar and, therefore, soluble in water. It acts as a weak base.
- **The sulfhydryl group** (-SH) helps stabilize the structure of proteins that have sulfur in their structure.
- **The phosphate group** is a dissociated form of phosphoric acid and is polar. Organic phosphates are very important in the cellular storage and transfer of adenosine triphosphate (ATP), a compound that releases free energy when its phosphate bonds are hydrolyzed.
- **The methyl group** (-CH_3) is a nonpolar, hydrophobic functional group. It contributes to the one of the principal structures of proteins.

IMPORTANT CELLULAR MACROMOLECULES

So far, we have described molecules as "tiny." That isn't always the case. Often organic molecules are large structures, or macromolecules (polymers). They are formed from simpler molecules called monomers. For example, carbohydrates (polymers) form simple sugars (monomers), and proteins (polymers) form amino acids (monomers). Structural variation of macromolecules is the basis for the enormous diversity of life. The theme of unity introduced in chapter one is reflected in the forty or fifty common monomers used to construct macromolecules. Think of each monomer as a letter in the alphabet of life and the macromolecules as the words. As the universal monomers are arranged in different ways, new properties emerge. The language of life develops and the theme is diversity.

Carbohydrates

Do you know anyone who has tried to lose weight by cutting carbohydrates from their diet? You can't avoid them anyway, since they are in almost every type of food you eat. Carbohydrates are the fuel and building material of organisms, and in food they are a major source of quick energy. The number of simple sugars present in the compound classifies them.

Monosaccharides are simple sugars. Glucose is the major nutrient of cells and the most common monosaccharide. Green plants produce glucose through photosynthesis. In fact, most of the food you eat is broken down into glucose. It is carried via the bloodstream to every cell in the body (See Figure 2.4). Energy stored in the chemical bonds of glucose is harvested during cellular respiration. It can form larger carbohydrates by condensation. The size of the carbon skeleton of carbohydrates varies from three or more, but the most common have three, five, or six carbons. The spatial arrangement around the asymmetric carbons may vary. In aqueous solutions, many monosaccharides form rings. Another typical monosaccharide is fructose, the sugar in fruit.

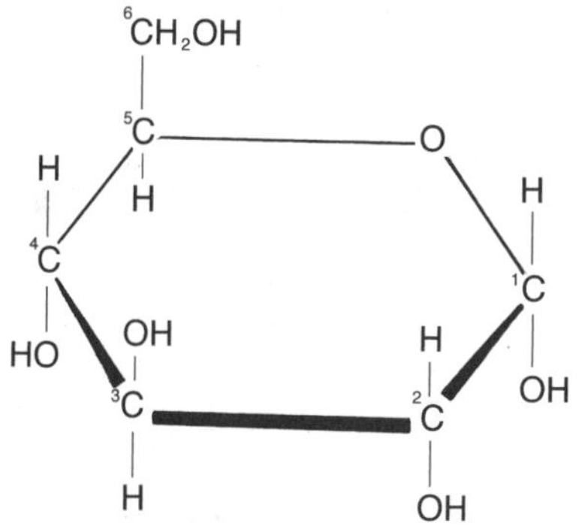

Figure 2.4—The Structure of Glucose

> Disaccharides form when two monosaccharides join through a process called dehydration synthesis (removal of water). Glucose and galactose (a simple sugar) combine to form the lactose found in milk. Glucose and fructose combine to form sucrose, or table sugar.

Nutrients that are not used as fuel are stored, some as fat and some as polysaccharides (long chains of simple sugars). Animals store polysaccharides as glycogen in the muscle and liver cells; this substance can be rapidly released when energy is needed.

Plants store polysaccharides in the form of cellulose in their roots, seeds, fruits, and tubers. Because this substance makes cell walls more rigid, it is called a *structural polysaccharide*. Cellulose is a form of dietary fiber for plant-eating humans. Because humans can't digest cellulose, it passes through the digestive tract.

- **Chitin is another structural polysaccharide. It forms the exoskeleton of the arthropods and is the building material in the cell walls of some fungi.**

Lipids

Lipids (commonly known as fats) are an essential part of a balanced diet, and they play a critical role in the body. Like carbohydrates, lipids can be used to store energy. When lipids are broken down, they produce more energy, gram for gram, than carbohydrates do. Triglycerides are the main form of lipid used to store energy in the body.

The formation of a triglyceride involves the combination of three fatty acid molecules with one glycerol molecule (See Figure 2.5). If every carbon atom in a fatty acid chain is joined to another carbon atom by a single bond, the fatty acid is said to be saturated. If a pair of carbon atoms is joined by a double bond, the fatty acid is said to be unsaturated.

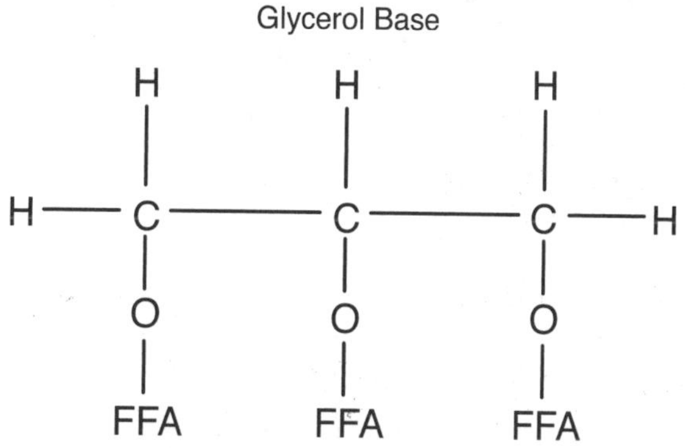

Glycerol Base

Figure 2.5 — Structure of a Lipid

> When you take in more glucose than the body can use or store, the leftovers are converted to triglycerides, or fat molecules. These travel through the bloodstream until they are stored in cells of adipose tissue.

Another example of lipids are phospholipids, which contain a phosphate group plus two fatty acids bonded to glycerol (see Figure 3.3). The head of the molecule is hydrophilic (attracted to water) and the fatty acid tails are hydrophobic (repelled by water). Steroids are lipids whose carbon skeleton has four fused rings (See Figure 2.6). An example is cholesterol, a common component of animal cell membranes. Cholesterol is a common precursor to many other steroids, including vertebrate sex hormones and bile acids.

Proteins

Proteins make up fifty percent or more of a cell's dry weight. They have a wide range of important and varied functions, which include:

- structural support (collagen)
- storage (of amino acids)
- transport (hemoglobin)
- signaling (chemical messengers)
- cellular response to chemical stimuli (receptor proteins)
- movement (contractile proteins)
- defense against foreign substances and disease-causing organisms (antibodies)
- catalysis of biochemical reaction (enzymes)

Proteins are polymers of amino acids arranged in a specific linear sequence and are linked by peptide bonds (See Figure 2.7). The function of the protein is dictated by the arrangement of the amino acids.

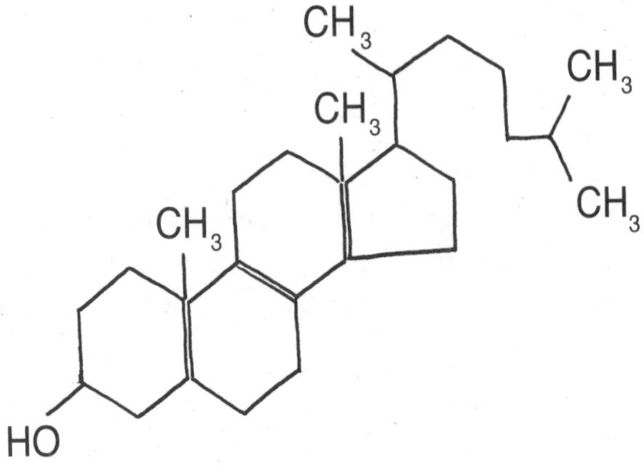

(a) Cholesterol

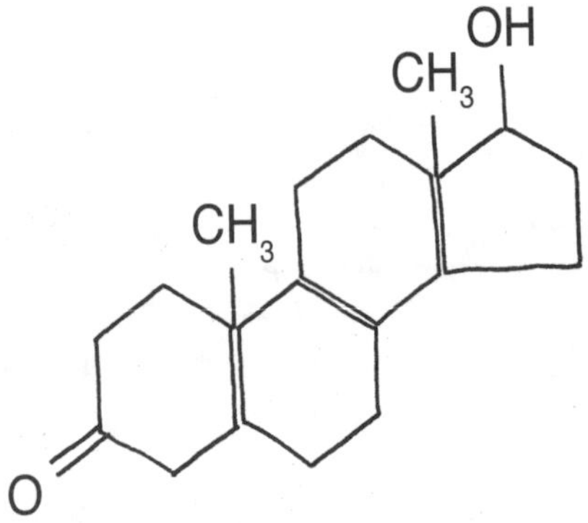

(b) Testosterone

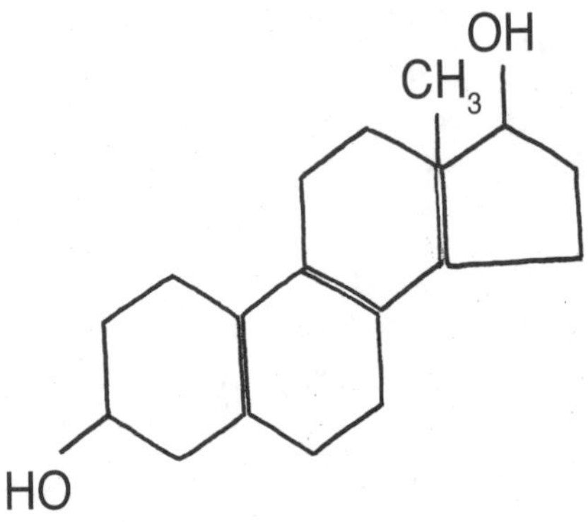

(c) Estradiol

Figure 2.6—Structure of Steroids

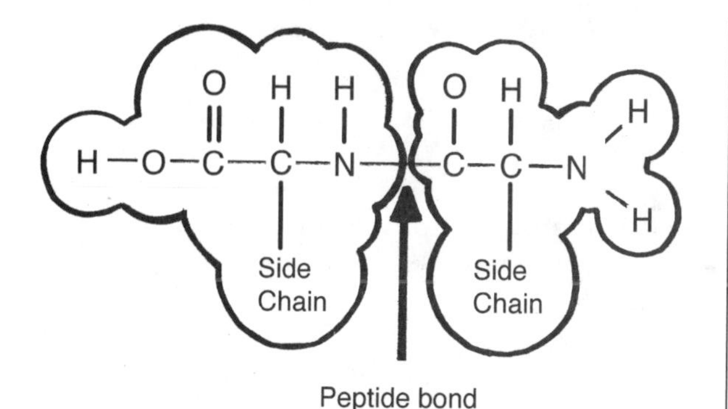

Figure 2.7—Structure of a Dipeptide

Nucleic Acids

Nucleic acids store and transmit the genetic information that is responsible for life itself. Nucleic acids are polymers of individual monomers known as nucleotides. The nucleotides are molecules built up from three parts: a five-carbon sugar, a phosphate group, and a nitrogenous base.

Deoxyribonucleic acid (DNA), a double-stranded helix, contains the sugar, deoxyribose (See Figure 2.8). Ribonucleic acid (RNA),

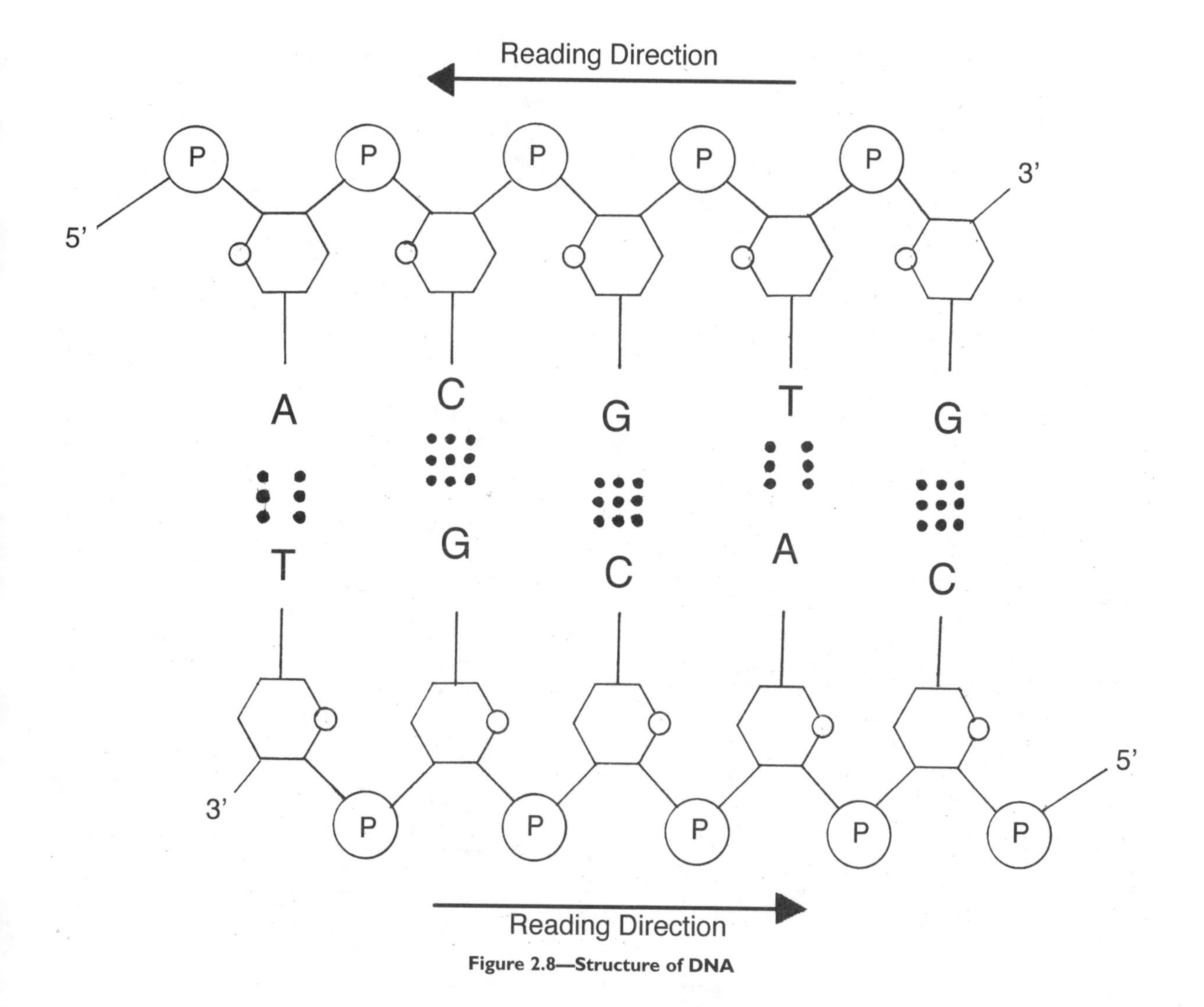

Figure 2.8—Structure of DNA

a single-stranded molecule, contains the sugar, ribose. Both DNA and RNA contain three nitrogenous bases: adenine, guanine, and cytosine. In addition, DNA contains thymine while RNA has uracil as its fourth nitrogenous base. The replication of DNA, transcription, and translation will be discussed in Chapter 7.

SUMMARY

- Chemical compounds form by the interaction of individual atoms where electrons can be shared, gained, or lost.
- The four most abundant elements in living organisms are carbon, hydrogen, oxygen, and nitrogen.
- Water is the most abundant compound in the majority of living things. It has several unique physical properties due to the formation of hydrogen bonds among the water molecules: versatility as a solvent, cohesiveness, imbibition, ability to stabilize temperature, and expansion upon freezing.
- Most biochemical reactions occur when the pH is between pH 6.0 and 8.0. Buffers are necessary to keep the pH from fluctuating.
- The major macromolecules of organisms are carbohydrates, lipids, proteins, and nucleic acids.
- Carbohydrates are the major fuel and building material of living things.
- Lipids can be used as a store for energy, components of plasma membranes or steroid hormones.
- Proteins play major roles in organisms including structural support, storage, transport, signaling, cellular response to chemical stimuli, movement, defense, and enzymatic reactions.
- Nucleic acids are responsible for storing and transmitting genetic information.

THE LIVING CELL AND ITS COMPONENTS

KEY TERMS

membrane	cytoplasm	diffusion
organelle	matrix	

Cells are the basic units of life. When working properly, cells perform the essential functions of life. Some living things are made of a single cell, while others are made up of trillions of cells. Some cells last only for a matter of hours (some white blood cells), while others can last a lifetime (brain cells). According to the *cell theory*, all types of organisms are composed of one or more cells. The cell theory also states that new cells must come from preexisting ones.

SIZE OF CELLS

The cell is the smallest functioning unit of a living thing that still has the characteristics of the whole organism. How small can a cell get? This is limited to the volume capable of holding genetic material, proteins, etc., which are necessary to carry out the basic cell functions and reproduction.

- **The smallest known cells are the mycoplasmas, which range in size from 0.1–10 μm.**

How large can a cell get? This is limited by metabolism. A cell must take in adequate amounts of oxygen and nutrients and get rid of wastes (carbon dioxide, for example). What happens to the relationship between the surface area (SA) and the volume (V) as a cell increases in size? Both are related to the radius of the cell as shown in the following equations:

$$SA = 4\ \pi r^2$$

$$V = 4/3\ \pi r^3$$

For small cells, the SA:V ratio increases, and for larger cells of the same shape, the SA:V ratio decreases. The primary mechanism of exchange of matter between a cell and its environment is diffusion (the slow mixing of molecules by random motion). The bigger the cell, the greater the distances the molecules have to travel. Surface area is the limiting variable. Cells of the same size can change their surface area by changing their shape. For example, to increase surface area without increasing volume, microvilli, the fingerlike projections of cells in the small intestine, change their shape.

TYPES OF CELLS

There are two basic types of cells, *prokaryotes* and *eukaryotes*. The prokaryotic cell has a distinct structure (See Figure 3.1). Bacteria and archaebacteria are of this type of cell. Shapes vary, but prokaryotic cells are bordered by a plasma membrane, which is surrounded by an external cell wall. Prokaryotes don't have a true nucleus. DNA is in a "nucleoid" region. With the exception of archaebacteria, proteins are not associated with bacterial DNA.

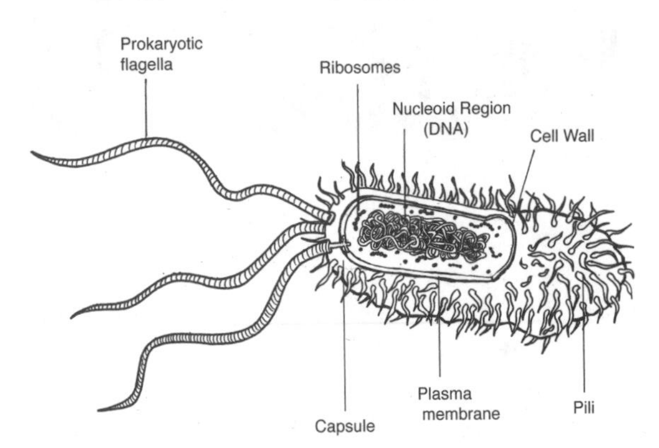

Figure 3.1—Typical Rod-Shaped Prokaryotic Cell

These cells are smaller than eukaryotic cells. They have no membrane-bound organelles. Enzymes necessary for certain metabolic activities are in the plasma membrane. Some of the cells are motile and have one or more flagella. We'll investigate bacteria and archaebacteria further in Chapter 9.

Protists (bacteria, protozoa, and algae) exhibited the earliest eukaryotic cells. Fungi followed suit, with plants and animals crowding onto the evolutionary bandwagon. In general, eukaryotic cells are larger than their prokaryotic counterparts. They have a distinct nucleus and an array of membrane-bound organelles on the cellular roster. We'll look into plant cells later in the chapter. Now let's explore the structures and functions of a typical animal cell (See Figure 3.2).

STRUCTURE OF CELLS

Between the cell and its environment is a semi-permeable barrier called the plasma membrane (See Figure 3.3).

Structurally, the plasma membrane is a phospholipid bilayer. Fatty acid "tails" meet at the center of the bilayer, repelling water. Polar "heads" of the phospholipids line the

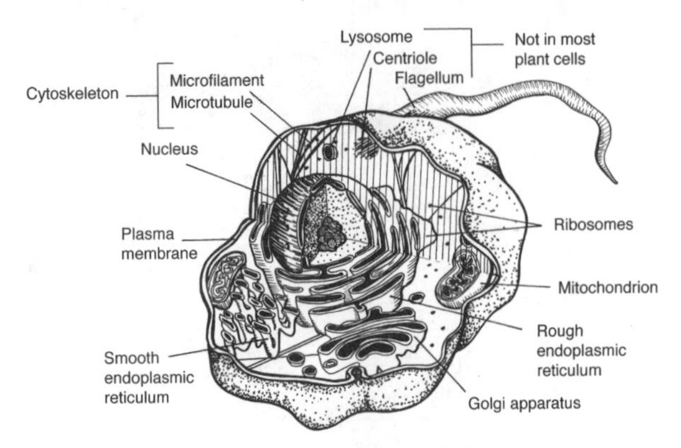

Figure 3.2—Animal Cell

top and bottom of the layer. These "heads" are attracted to water. Cholesterol (in animal cells) keeps the bilayer fairly fluid at body temperature, more like thick oil than fat.

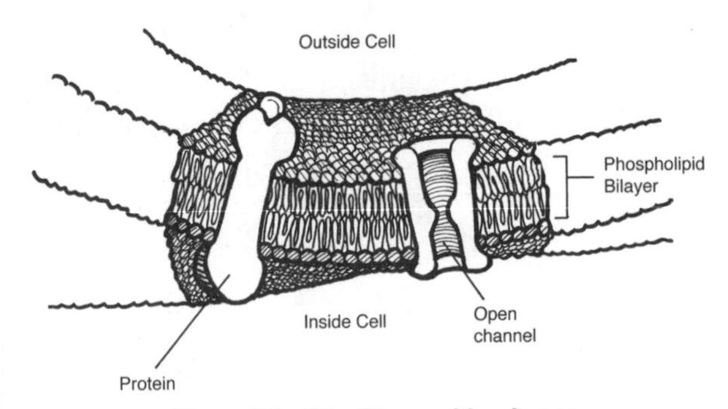

Figure 3.3—The Plasma Membrane

Proteins are embedded in the membrane. Some pass all the way through. Others go only part of the way, while still others are peripheral. Some are fixed in position. Some move laterally.

- **The prefix cyto means cell.**

The membrane regulates what enters and leaves the cells. The size, hydrophobicity, polarity and charge of a molecule will determine if it can cross the membrane, and if so, how. Keep in mind that the core of the plasma membrane is hydrophobic—repels water.

Gases such as oxygen and carbon dioxide, small uncharged polar molecules, such as water, urea, and ethanol, and hydrocarbons can go through these membranes. Some molecules must pass through open channels in the proteins. Carbohydrates at the surface of the membrane act as receptors, allowing molecules that they recognize to pass and refusing entry to others.

When large uncharged polar molecules like glucose, and large charged molecules, such as amino acids, must cross the plasma membrane, proteins provide the answer. Some act as carriers. If they move down a concentration gradient, it does not require ATP (see page 25) and is called facilitated diffusion.

Diffusion is a form of passive transport. Molecules simply move across the membrane from areas of high concentration to areas of low concentration. It ceases when concentrations are evenly distributed. Oxygen and carbon dioxide use this method. Water moves across by a similar process called osmosis.

Sometimes molecules are moved against their concentration gradient. This takes energy on the part of the cell. We call this active transport, which usually involves ATP.

Nucleus

The nucleus is the control center of the cell. Through instructions carried in DNA, it tells the cell what to do. It is about 2–5 μm in diameter and is surrounded by a double membrane called the nuclear envelope. The DNA in the nucleus forms large structures called chromosomes, which are attached to proteins, called histones. The processes of DNA replication and transcription take place in the nucleus. RNA leaves the nucleus via pores in the nuclear envelope, and travels to the cytoplasm, the cell interior.

Nucleolus

Most nuclei contain a small region called the nucleolus, made up of RNA and proteins. The ribosomes, particles used to synthesize proteins, form here, exit via pores in the nuclear envelope, and enter the cytoplasm, where they are involved in protein synthesis.

Endoplasmic Reticulum

Many cells are filled with a complex network of flattened tubular structures known as the endoplasmic reticulum (ER). The membrane of the ER is continuous with that of the nucleus. The ER transports material through the inside of the cell. There are two distinct regions of endoplasmic reticulum that differ in structure and function.

The rough ER is dotted with protein-producing ribosomes, hence the name rough. The protein goes through the internal channel of the ER and folds into its own conformation. The ER surrounds the protein with a sac, called a vesicle, pinched off from its own membrane. The next stop is the "cis" area of the Golgi apparatus, where it can be modified before leaving the cell. Proteins destined to leave a cell (secretory proteins, such as insulin) are the ones transported by the rough ER.

- **Ribosomes are involved in building protein from amino acids. They usually exist in two parts, or sub-units, that come together when it's time to work.**

The smooth ER is involved in the synthesis of lipids, including phospholipids and steroids, such as steroid hormones produced by the adrenal glands, and the sex hormones. Not surprisingly, the ovaries and the testes are rich in smooth ER. In muscle cells, the membrane of the smooth ER pumps calcium necessary for muscle contraction.

The smooth ER has a role in the metabolism of carbohydrates and the detoxification of drugs and poisons. Because of this it is abundant in the liver, the main site of detoxification. Glycogen is stored in the liver and the smooth ER produces enzymes necessary in the conversion of glycogen to glucose. A hydroxyl group is often added, making the substance more soluble in water, and so, more easily flushed out of the body.

Golgi Apparatus

The Golgi apparatus looks like a stack of flattened pancakes. This organelle is a continuation of the ER membranes. It stores, sorts, and transports the ER products to their final destination outside of the cell. It is a site of protein modification (e.g., additions of carbohydrates). The Golgi apparatus has a structural polarity. The "cis" side receives the ER product, and the modified protein leaves the Golgi via the "trans" side.

- **Italian scientist Camillo Golgi first described the Golgi apparatus in 1898.**

Lysosomes

Lysosomes are membrane-bounded sacs of hydrolytic enzymes. An array of enzymes—carbohydrases, protease, lipases, and nucleases—enable lysosomes to digest all major classes of macromolecules. The membrane segregates potentially destructive enzymes from the cytoplasm. It maintains an optimal acidic environment (pH 5.0) for enzyme activity by pumping in hydrogen ions.

- **If the lysosome membrane ruptures, the enzymes inside can spread throughout the cell and destroy it. The name lysosome means "bodies that dissolve."**

The lysosomes have several important functions. They may fuse with food-filled vacuoles and employ enzymes to digest the food. They recycle the cell's own organic material, digesting the contents, causing monomers to be released and used as new organic building blocks. Lysosomes are responsible for programmed cell destruction, which is important during development and metamorphosis.

In the genetic disorder, Tay-Sachs disease, lysosomal lipase is missing or inactive. This causes lipid accumulation in the brain and death at a young age.

Mitochondria

The sandwich you ate for lunch is destined to provide fuel to a tiny bean-shaped organelle called a mitochondrium. The mitochondria convert the fuel to useable energy. This organelle contains its own DNA and ribosomes. It is semiautonomous, and replicates itself.

Mitochondria are enclosed by a double membrane. The smooth outer membrane is highly permeable to small solutes and blocks the passage of proteins and other macromolecules. The inner membrane has infoldings called the cristae, which has enzymes involved in cellular respiration embedded in it. The inner folding increases surface area. Nature often makes use of this strategy. The matrix of the mitochondria contains enzymes that catalyze many of the metabolic steps of cellular respiration.

- **A single cell in the human liver may have up to 1000 mitochondria. An average number per cell is about 200.**

Peroxisomes

The peroxisome is an organelle found in nearly all eukaryotes and it is bound by a single membrane. It contains enzymes that transfer hydrogen to oxygen, thus producing peroxide (which is toxic). Peroxide is also produced in some metabolic reactions. Catalase then breaks down the peroxide to water and oxygen. This is useful in removing a hydrogen from alcohol in the liver cells. Peroxisome enzymes also are involved in the breakdown of fatty acids to acetyl CoA, which is transported to the mitochondria for fuel during cellular respiration.

Vacuoles

Vacuoles are membrane-enclosed sacs with assorted functions in cell maintenance. Food vacuoles are formed by phagocytosis, the process in which white blood cells consume particles. Some fresh-water protozoa contain contractile vacuoles, which pump excess water from a cell. Most mature plant cells contain a long central vacuole (see page 22).

Cytoskeleton

There are three types of cytoskeleton fibers: microtubules, microfilaments, and intermediate filaments. Microtubules are in all eukaryotic cells. They are straight, hollow fibers made from globular proteins called tubulin. They have many functions including cellular support, organelle movement, cell motility, and the separation of chromosomes during cell division.

Microfilaments are made up of two intertwined strands of the protein, actin. They provide cellular support when they combine with other proteins just inside of the plasma membrane and play a role in cell shape. They are the core of the microvilli in the epithelium of the intestinal wall. They participate in muscle contraction and localized contraction of cells. During cell division of an animal cell, the microfilaments form a contractile ring, which pinches the animal cell in two. In plants, they are involved in cytoplasmic streaming (movement within the cell) and they have a hand (or should we say a foot) in the elongation and contraction of pseudopods that power an amoeba from place to place.

Intermediate filaments are composed of keratin subunits and are more permanent than either microtubules or microfilaments. They may be the framework of the cytoskeleton. They reinforce cell shape and probably fix an organelle's position in the cell. These fibers line the interior of the nuclear envelope.

Intercellular Junctions in Animal Cells

Virtually all cells are in contact with neighboring cells. The way that they interact with their neighbors varies. There are at least three different types of contacts: gap junctions, desmosomes, and tight junctions.

- **Gap junctions** are like protein doorways between adjacent cells. They allow the flow of materials and electrical charge.
- **Desmosomes** are strong anchors between neighboring cells. They are interacting complementary folds of membrane. Put

your hands together by interlinking your fingers and you'll get the basic idea.

- **Tight junctions** are bands of proteins that prevent fluids and small molecules from crossing the membrane. Tight junctions in the stomach lining protect the stomach cells from hydrochloric acid.

Cilia and Flagella

Some cells are motile—they move. In particular, single-celled organisms must move to find food. Even some mammalian cells, such as sperm cells, move independently. Some cells form linings that create movement across their surface. What facilitates this movement? There are two basic structures involved. Many freely moving cells have a long tail-like structure called a flagella. Other cells move via cilia, a series of hair-like fibers that beat in unison. Cilia and flagella are formed from a core of nine outer microtubules and two inner single microtubules ensheathed in an extension of the plasma membrane.

Extracellular Matrix

Animal cells do not have a cell wall, but they do have an extracellular matrix (ECM)—an external network of protein/carbohydrates that supports the cell, and functions in cell adhesion and attachment, development, and movement. The most abundant glycoprotein in the ECM of most animal cells is collagen, which forms strong fibers outside of the cells. Other glycoproteins attach the ECM to the plasma membrane.

Unique Plant Cell Structures

Plant cells are encased by a cell wall, which is thicker than the plasma membrane (See Figure 3.4). The composition of the cell wall varies, but there are strong cellulose fibers embedded in a matrix of other polysaccharides and proteins. It protects plant cells, maintains their shape, prevents excess water uptake, and has membrane-lined channels, plasmodesmata, which connect the cytoplasm of neighboring cells.

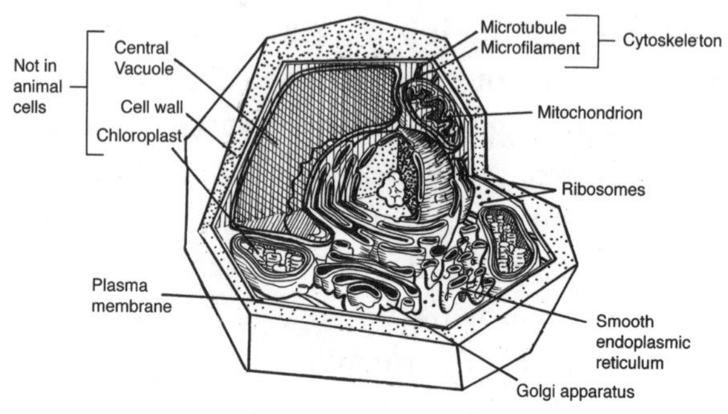

Figure 3.4—Plant Cell

A young plant cell secretes a thin, flexible primary cell wall. Between the primary cell wall of adjacent cells is a middle lamella (thin membrane) made up of pectins, a sticky polysaccharide that cements cells together. As cells stop growing, the cell wall is strengthened. Some of the plant cells secrete hardening substances into the primary cell wall. Others add a secondary cell wall between the plasma membrane and the primary wall. The secondary cell wall is often deposited in layers with a durable matrix that supports and protects the cell.

- **Tiny holes in the cell walls of plants allow the passage of molecules. These holes are called plasmodesmata—which means "little holes in the cell wall."**

The central vacuole is a large, fluid-filled organelle of a mature, living plant cell. It is enclosed by a membrane called the tonoplast and has many functions, depending upon the species. It can store organic compounds (protein storage in seeds), store inorganic ions, keep dangerous metabolic by-products from the cytoplasm, contain soluble pigments (red or blue pigments in flowers), or protect the plant by containing poisonous or distasteful compounds. As it enlarges during growth, it forces the primary cell wall to expand and cell surface to increase.

The rosy color of an apple, the golden hues of autumn leaves, and photosynthesis have something in common—plastids. Plastids are a group of membrane-bound organelles. Amyloplasts are colorless and store starch in roots and tubers. Chromoplasts contain pigments other than chlorophyll and contribute to the color of fruits, flowers, and the autumn leaves. Chloroplasts contain the pigment, chlorophyll, and are the site of photosynthesis in plants. Chloroplasts, lens-shaped organelles, contain their own DNA and can replicate autonomously.

SUMMARY

- All living organisms are composed of one or more cells. All cells have a plasma membrane, DNA, and cytoplasm.
- Cells are divided into two categories based on their structure: prokaryotic or eukaryotic. Bacteria and archaebacteria are prokaryotic cells. They are smaller, lack membrane-bound organelles, and have DNA in a nucleoid region. Within the cytoplasm of eukaryotic cells are organelles, which perform specialized functions. Protists, fungi, plants, and animals have a eukaryotic cell.
- Animal cell organelles include nucleus, nucleolus, endoplasmic reticulum, Golgi apparatus, lysosomes, mitochondria, peroxisomes, vacuoles, and cytoskeleton. If the cell is motile, it may contain cilia and/or flagella. Communication between animal cells is accomplished by gap junctions, desmosomes, or tight junctions.
- Plant cells are protected by a cellulose-containing cell wall. They have several structures unique to them. The large central vacuole takes up most of the space of a mature, living plant cell and has varied functions.
- Plant cells are equipped with chloroplasts, which contain the pigment chlorophyll, necessary for the conversion of light energy to chemical energy in the process of photosynthesis.

ENERGY AND METABOLISM

KEY TERMS

respiration	pyruvate	catabolic pathway
enzyme	phosphorylation	anabolic pathway
catalyst		

Energy, the ability to do work, comes in many forms. The energy required for motion is called *kinetic* energy. Light and heat are forms of kinetic energy. Stored energy is called *potential* energy. The gasoline in your car contains potential energy that your engine can convert into kinetic energy. In living cells, we find both kinetic and potential energy. Kinetic is the energy of pressure and movement, whereas potential is the energy stored up in the fuels our cells need to function and do work.

LAWS OF THERMODYNAMICS

Energy can be converted from one form to another, and thermodynamics is the study of these changes. Energy can neither be created nor destroyed, and thus the amount in the universe is constant. This is the first law of thermodynamics.

The second law of thermodynamics tells us no energy conversion is perfect. With every conversion, some of the energy always winds up as heat. In addition, every energy conversion increases the disorder and disorganization of matter (entropy). This principle has direct application to living cells, because cells cannot transfer or transform energy with one hundred percent efficiency due to the second law of thermodynamics. When you exercise, you contract your muscles to produce motion, but heat is always a by-product of this movement. All of the heat used to maintain our body temperature comes from chemical reactions.

In case you are curious, the third law of thermodynamics deals with thermal equilibrium. If A is in equilibrium with B, and B is in equilibrium with C, then A is in equilibrium with C. It's called enthalpy.

ROLE OF ADENOSINE TRIPHOSPHATE

Adenosine triphosphate (ATP) is a source of potential chemical energy for most enzyme reactions. It is composed of adenine (a nitrogenous base), ribose (a five-carbon sugar), and a chain of three phosphate groups covalently linked to ribose. The last two bonds are not stable but can store up chemical energy, much as a battery stores electrical energy. So, when energy is needed to drive a chemical reaction in the body, this energy can be harvested from the stored energy in ATP. To release ATP's energy, the last phosphate group is removed. This removal releases energy, liberates a free phosphate group, and leaves behind ADP, adenosine diphosphate. Specialized proteins, called enzymes, use the energy released from ATP to do all the work of the cell. This includes the manufacture of new molecules, pumping of ions in or out of the cell, etc. Often, this work involves the transfer of the terminal phosphate group of ATP to another protein in a process called phosphorylation. This is a key regulatory process in every cell of the body. Given all the important jobs that ATP has to do, it should be no surprise that cells, mostly in the mitochondria, are constantly generating ATP. But where does the chemical energy needed to make ATP come from? It comes from food that has been eaten and digested.

- **A cell at work uses and regenerates its total supply of ATP about once every sixty seconds.**

ENZYMES: BIOLOGICAL CATALYSTS

All chemical reactions are really just about the transfer of energy from one molecule to another. Almost all of this energy is stored in the bonds that hold the atoms of a molecule together. When a molecule's energy is changed, bond breaking and bond forming are always involved. Often, to get a bond to change, we must inject a little energy to "prime" the bond and get it to an excited state where it's ready to change. This is a little like starting a siphon—we have to first get the fluid up over an activation barrier, after which it will flow downhill. The amount of energy that reactant molecules must absorb to start a reaction is called the activation energy. Sometimes activation energy is affected by catalysts, which facilitate the organization and distribution of molecular energy in a way that accelerates a reaction without permanently changing the catalyst. In living systems, enzymes catalyze most reactions. By helping to bring together reactant molecules, enzymes lower the activation-energy threshold, making it possible for reactions to occur at cellular temperatures. As catalysts, enzymes do not change the chemical nature of a reaction, but simply speed it up. Another great advantage of enzymes is that they are very selective and reactant.

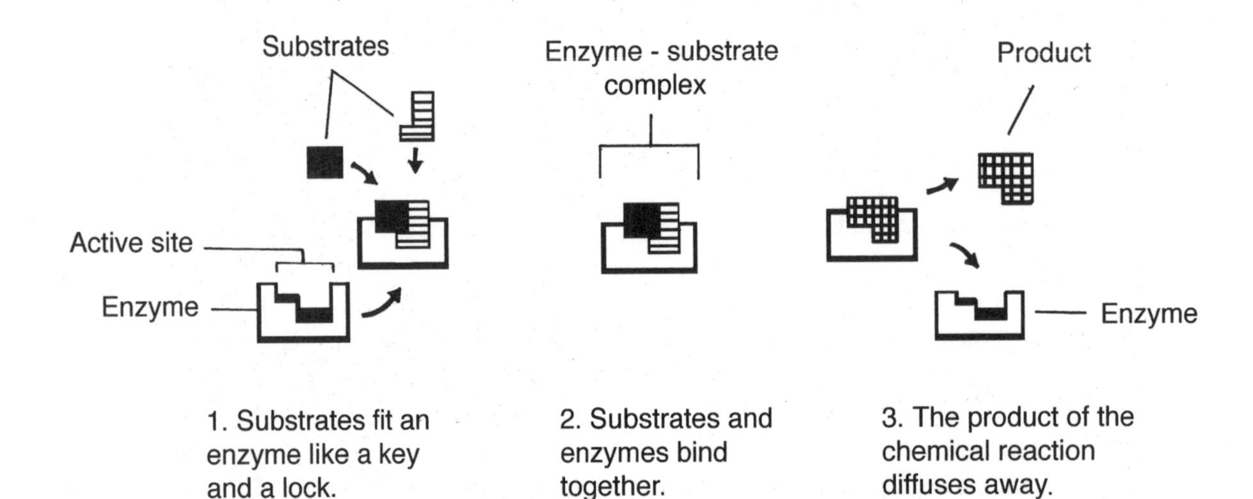

Figure 4.1—How Enzymes Work

Enzyme and substrates (the substances acted upon) must bind tightly together at the active site, like a key in a lock (See Figure 4.1). The active site may hold two or more reactants in the proper position, which allows them to react and form a product. In other cases, the enzyme's active site distorts the chemical bonds of the substrate, and less energy is needed to break them during a reaction. In some cases, the active site creates a microenvironment, such as changing the local hydrogen ion concentration caused by nearby basic amino acids. In other cases, the amino acids at the active site may play a direct role in the reaction.

Several conditions influence the rate of a reaction. Optimum pH and temperature conditions must be met for maximum enzymatic activity.

- **Most enzymes work best between pH 6 and pH 8 and at temperatures between 35°–40° C.**

These optimum conditions allow the greatest number of molecular collisions without denaturing or unfolding the enzyme. In general, the velocity of an enzymatic reaction increases with increasing temperatures because the kinetic energy of the molecules increases the substrate collisions with the active sites of the enzyme. Reactions that occur above the optimum temperature may decrease the reaction rate if the protein becomes denatured. Heat disrupts the weak bonds that stabilize a protein's shape.

Some enzymes need cofactors for proper activity. Cofactors are nonprotein helpers that function as catalysts. They may be inorganic (e.g., zinc, copper, or iron) or organic (vitamins, for example). Some bind tightly to the active site, while others bind loosely to both the active site of the enzyme and the substrate.

The greater the substrate concentration, the faster the reaction—up to a limit. If the substrate is present in high enough quantities, the enzyme becomes saturated (no more available binding sites). When the enzyme is saturated, the reaction rate depends upon how fast the enzyme can complete each reaction cycle. When an enzyme is saturated, adding more enzyme will increase the reaction rate.

Metabolic pathways are chains of enzymatic reactions, one linked to another. Controlling enzyme activity at various steps along the pathway regulates the flow of energy through these pathways. One way to control enzyme activity is to control the amount of enzyme present, which is done by turning on and off the genes needed to synthesize each enzyme. Another way to turn enzymes on and off is to add chemical switches called inhibitors or activators in specialized chemical reactions. For example, the addtion of a phosphate molecule through phosphorylation regulates many different enzymes. The important point is that enzyme activity is carefully regulated by numerous mechanisms.

CELLULAR RESPIRATION

Cellular respiration is the process of breaking down food molecules and capturing their energy in ATP molecules. This breakdown process releases high-energy electrons that ultimately travel to the mitochondria where they combine with oxygen. This breakdown process is organized into several steps, all of which are controlled by enzymes. In one step, glucose is slowly oxidized in a process called glycolysis to yield a few high-energy electrons and a small molecule named pyruvate. More high-energy electrons are harvested from pyruvate in the Krebs cycle (see page 28).

All of these high-energy electrons are ultimately sent to the electron transport chain inside the mitochondria to fuel the synthesis of ATP. At every step of this pathway known as cellular respiration, there is a very close association of structure and function.

Glycolysis

Glycolysis is a series of reactions that take place in the cytoplasm of all cells. It does not require oxygen. In glycolysis, glucose (a six-carbon sugar) is partially oxidized and rearranged to form two three-carbon pyruvate molecules.

There are two phases in glycolysis. In the first phase, energy in the form of ATP is added. The main compound formed in this reaction is glyceraldehyde-3-phosphate, which is then broken down into two molecules of pyruvate with the simultaneous formation of two molecules of ATP and two molecules of an energy-carrying molecule named nicotinamide adenine dinucleotide (NADH). The low energy form of NADH is NAD+, an electron acceptor that shuttles electrons and hydrogens from glycolysis to the mitochondria. Most of the energy harnessed in glycolysis is conserved in the high-energy electrons of NADH and ultimately in the phosphate bonds of ATP. The fate of pyruvate depends upon the presence or absence of oxygen. If oxygen is present, pyruvate is converted into a molecule named acetyl CoA and then goes on to enter the Krebs cycle. If oxygen is not present, the pyruvate is converted to lactic acid, which builds up until the oxygen supply is restored.

- **The process of conversion of pyruvate into lactate is called fermentation.**

The Krebs Cycle

The Krebs cycle (citric acid cycle) takes place in the mitochondrial matrix (See Figure 4.2). The acetyl-CoA sent from glycolysis combines with the compound oxaloacetate, and then enters the Krebs cycle. This cycle is a series of enzymatic reactions that ultimately break down the original carbons from pyruvate into carbon dioxide with the parallel synthesis of two molecules of ATP, six molecules of NADH, and four molecules of $FADH_2$ (another high-energy electron carrier). It takes two turns of the Krebs cycle to completely oxidize glucose. At the end of the cycle, oxaloacetate is regenerated. Compared to glycolysis, the Krebs cycle yields much more energy in the form of NADH and $FADH_2$, both energy-rich molecules. In spite of its increased energy production, the Krebs cycle is still not enough to power all of an organism's activities.

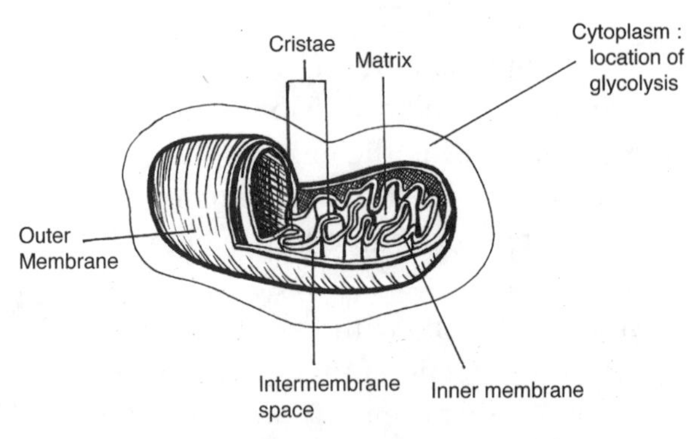

Figure 4.2—Structure of a Mitochondrion

Electron Transport Chain and Chemiosmosis

The electron transport chain does not produce ATP directly. Instead, it generates a proton (H^+) gradient across the inner mitochondrial membrane, which stores the potential energy used to phosporylate ADP to yield ATP (chemiosmosis). Most of the electron carriers of the electron transport chain (ETC) are proteins embedded in the inner mitochondrial membrane (cristae). A notable exception is ubiquinone, which is a lipid.

Each successive carrier molecule is more electronegative than the previous one. Oxygen (with the greatest electronegativity) is the final electron acceptor. Multiple copies of the protein complex ATP synthase are found in the mitochondrial cristae. It is this enzyme complex that makes the bulk of cellular ATP. It uses the proton gradient across the inner mitochondrial membrane to power ATP synthesis. Some of the electron carriers of the ETC carry only electrons and others accept and release protons along with the electrons. The protons picked up from the matrix are released into the intermembrane space. As electrons are shuttled between molecules, enough free energy is released to do work (See Figure 4.3). The ETC and chemiosmosis generate about ninety percent of the cell's ATP (34 molecules).

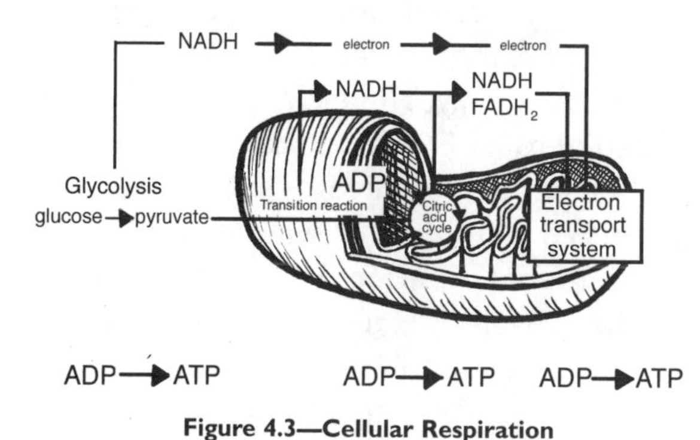

Figure 4.3—Cellular Respiration

Glucose is not our only cellular fuel. All the simple sugars in our diet are catabolized and each enters at its own point in cellular respiration (See Figure 4.4). Different sugars enter glycolysis at various steps along the pathway. Glycerol enters at the last step in the first phase of glycolysis, while fatty acids enter the transition stage between glycolysis and the

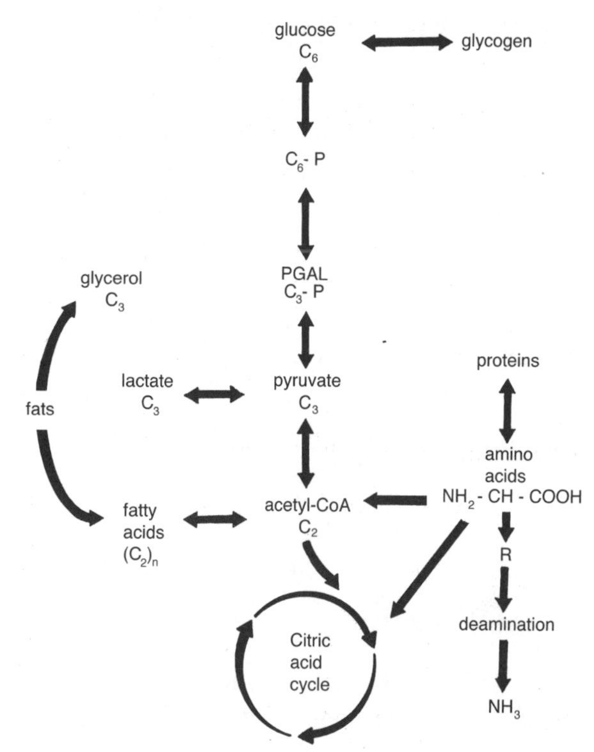

Figure 4.4—Macromolecule Energy Sources

Krebs cycle. Some amino acids are converted to pyruvate and others enter at the level of the Krebs cycle.

As with all complex metabolic pathways, cellular respiration is under enzymatic control. Cellular respiration is inhibited by large amounts of ATP and citrate (an early product of the Krebs cycle). It is stimulated by high amounts of AMP (adenosine monophosphate). In this way, the energy balance of the cell can be carefully monitored and regulated.

Pyruvate, the end product of glycolysis, is a key juncture in catabolism, the breakdown of large molecules to smaller ones resulting in energy release. In the presence of oxygen, it enters the Krebs cycle. In the absence of oxygen, it enters an anaerobic process called fermentation. The byproducts of this process are alcohol or lactate.

PHOTOSYNTHESIS

In the process of photosynthesis, an organism is able to produce carbohydrates and oxygen from water and carbon dioxide. In the light reaction, solar energy is used to make chemical energy, which powers the Calvin cycle (see below) in which carbon dioxide is incorporated into organic molecules. In essence, photosynthesis is the reverse of cellular respiration. The main end products of respiration are carbon dioxide (CO_2) and water, and these are used as the starting material for photosynthesis, which converts them into glucose and oxygen.

What Is Light?

Light is a form of electromagnetic energy. Visible light (wavelengths between 380 and 750 nm) is detected as colors. There is an inverse relationship between wavelength and energy; as the wavelength of light increases, the energy decreases and vice versa. When light meets matter, it can be reflected, transmitted, or absorbed. Pigments absorb visible light. While light is a mixture of all of the wavelengths of visible light, blue and red wavelengths are the most effectively absorbed by the plant pigment, chlorophyll.

Role of Chlorophyll

Plant cells contain a variety of pigments. Chlorophyll is a pigment that absorbs light energy. Types *a* and *b* are found in specialized structures called chloroplasts and give plants their green color. Red and blue wavelengths are absorbed and green is reflected. There is a structural difference between chlorophyll *a* and *b*, and they absorb different wavelengths of light. Chlorophyll *a* participates directly in the light reaction. Carotenoids (yellow, orange, red, and brown pigments) are also present in the plant cell and may absorb excessive light that would otherwise damage the chlorophyll. The site of photosynthesis is typically in the leaf of a green plant. Each cell has about thirty to forty chloroplasts.

- **Chlorophyll resides in the thylakoid membranes of the chloroplasts.**

Light-Dependent Reactions

The light reactions absorb the energy of sunlight and convert it to the energy that is stored in chemical bonds. The thylakoid membranes of the chloroplasts contain clusters of pigment molecules, or photosystems, that are able to absorb the energy in sunlight. In green plants, there are two systems: photosystem I and photosystem II. Each photosystem contains several hundred molecules of chlorophyll, as well as accessory pigments. Once the light is absorbed, the electrons of chlorophyll become excited. This energy is passed along until it reaches a particular pair of molecules, which can process this energy. These electrons are transferred to an ETC in the thylakoid membranes. A H^+ gradient is set up and ATP is produced using an ATP synthase complex (similar to that found in the mitochondria). This energy is needed to drive the Calvin cycle. In addition, $NADP^+$ is reduced to NADH, which is also sent to the Calvin cycle. As a result of the light reaction, water is split and oxygen is given off. The plant uses some of this oxygen for cellular respiration, and some is released into the atmosphere.

The Calvin Cycle

The series of chemical changes that make up the Calvin cycle is particularly critical to living things. In this part of photosynthesis (which occurs in the chloroplasts), the simple inorganic molecule of carbon dioxide is used to make a complex organic molecule. Glyceraldhyde-3-phosphate is the end product of the Calvin cycle. For every three carbon dioxide molecules that enter the plant (through the stomates in the leaf), nine molecules of ATP and six molecules of NAPDH are expended to produce one three-carbon molecule. The light reactions contribute the energy needed for the Calvin cycle. Only one sugar molecule exits to be used by the plant cell. It can be used to make other sugars and complex carbohydrates, fats, protein, or broken down to release energy. The rest of the glyceraldehyde is recycled in order to regenerate ribulose bisphosphate—the compound that begins the Calvin cycle and combines with carbon dioxide. Many plants make more sugar than they need. This sugar is converted to starch and is stored in the roots, tubers, and fruits of the plants.

C_3, C_4, and CAM Plants

Plants in which the Calvin cycle uses carbon directly are called C_3 plants because a three-carbon sugar is formed. C_3 plants—including most green plants—take in CO_2 during the day through their *stomata*, pores in their leaves and stems. C_4 plants like corn have special ways to save water when the weather is hot and dry: their stomata close during the day. The carbon is fixed into a four-carbon compound, which acts as a shuttle when it donates CO_2 to a nearby cell.

Another mode of carbon fixation and water conservation called CAM (Crassulacean acid metabolism) has evolved. Most succulent plants, cacti, and pineapples only open their stomata at night. The CO_2 is fixed into a four-carbon compound at night. During the day, carbon dioxide is removed from this four-carbon compound and enters the Calvin cycle.

SUMMARY

- Energy is the capacity to do work. Energy can be converted from one form to another, but it cannot be created or destroyed. Entropy in the universe increases during energy conversion.
- ATP is a molecule present in living organisms, which shuttles energy within cells. Enzymes are important biological catalysts. They function in lowering the activation energy needed for a reaction to occur at cell temperatures. Enzymes are neither changed nor used up in the reaction.
- Cellular respiration is a catabolic process in which energy stored in food molecules is transferred to the bonds of ATP. Glycolysis is the first step in this process and takes place in the cytoplasm. All organisms can carry on glycolysis. If oxygen is present, the food molecules are further catabolized during the Krebs cycle and then the electron transport chain. The most energy comes from chemiosmosis due to large amounts of ATP formed. The Krebs cycle takes place in the mitochondrial matrix and the ETC occurs in the cristae. If oxygen is not present, the final product of glycolysis, pyruvate, is fermented into ethanol or lactate.
- Most organisms need the food manufactured by green plants in the process of

photosynthesis. This is an anabolic process in which organic molecules are made from inorganic CO_2 and H_2O. Chlorophyll, the main light-absorbing molecule of green plants, is located in clusters in the chloroplasts. The light-dependent reactions occur in the thylakoid membranes of the chloroplasts. ATP and NADPH are generated during this process and become the energy needed for the Calvin cycle, which takes place in the stroma of the chloroplasts.

- Oxygen is a by-product of the Calvin cycle. Most plants are C_3 plants, which take carbon directly from the air and use it to build a three-carbon sugar. Modifications in the process of photosynthesis are seen in plants that inhabit hot, arid climates. Since their stomata are closed during the day to conserve water, they have an adaptation that allows them to form a four-carbon compound from CO_2. This CO_2 can be used during the day in the Calvin cycle or sent to another type of cell for the manufacturing of sugar. The stored sugars are a major source of food for many animals.

CHAPTER 5

THE CELL CYCLE

KEY TERMS

chromatid	chromosome	gamete
haploid	diploid	homologous

Approximately two trillion cell divisions occur in an adult human every twenty-four hours. That is twenty-five million divisions per second! The cells of your skin, hair follicles, bone marrow, and the lining of your gastrointestinal tract are turning over constantly. Cells divide for reproduction (in bacteria and amoeba, for example), to allow an organism to grow and develop (the fertilized egg), and to repair or replace damaged or dead cells. They can only divide when adequate cues are present: proper nutrients, growth factors, density, and volume.

- **An adult human has 2.5 trillion red blood cells, which have an average life span of 120 days. Stem cells in the bone marrow divide and produce 2.5 million red blood cells every second!**

INTERPHASE

Cells spend most of their time in interphase, which includes the three stages of the cell cycle, G_1, S, and G_2 (See Figure 5.1). They don't divide at this point, but carry on their normal cellular activity. The cell volume and number of organelles double and DNA replication gets underway.

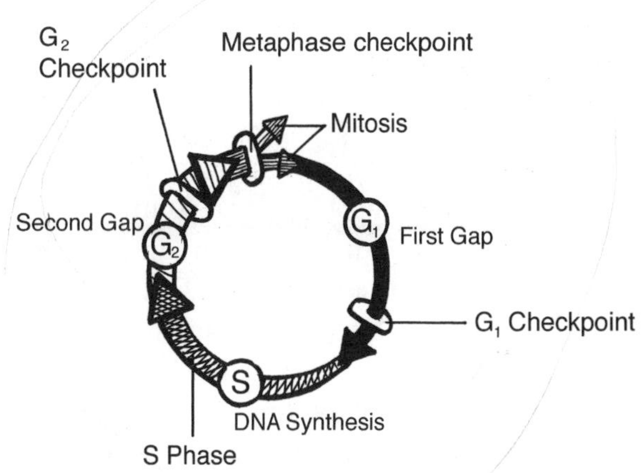

Figure 5.1—The Cell Cycle with Checkpoints

The G_1 phase, or gap phase, is a period of intense biochemical activity. Cytoplasmic structures and molecules increase in number. To house everything, the cell doubles in size. If centrioles (small granules outside the nuclear membrane) are present, they begin to replicate. This is the longest phase and some cells never leave it.

In the S phase, the cell is at work on one of life's most critical activities—the replication of DNA. While DNA is being copied exactly, DNA-associated proteins are synthesized.

G_2 is the final period of interphase. The nucleus is still well defined and surrounded by a nuclear envelope. Replication of the centrioles is completed. The spindle apparatus that helps to move chromosomes during mitosis begins to be assembled. The chromosomes begin to coil and condense. The cell is ready to leave interphase.

- **Cells that are in a non-dividing phase for a long time, such as liver and muscle cells, enter a phase called G_0. There is no replication of DNA in these cells.**

TYPES OF CELL DIVISION

Binary Fission

Bacteria divide by binary fission (See Figure 5.2). Prior to cell division, the single, circular chromosome replicates. The two strands move to opposite ends of the cell by an unknown method (there is no mitotic spindle). Once the cell doubles in size, the plasma membrane grows inward, laying down new cell wall material as it goes. When the new cell wall is in place, it "pinches" off. The cell divides into two, producing two genetically equivalent daughter cells. Binary fission is fast. Under ideal conditions, it can be completed in twenty to thirty minutes.

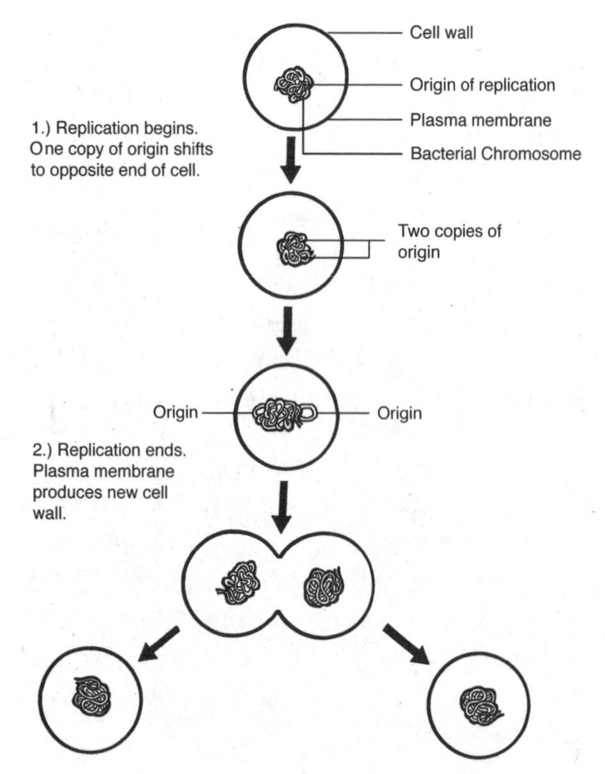

Figure 5.2—Binary Fission

Mitosis

Cell division is more complex in eukaryotic cells. The cells are much larger and more complex. Materials must be distributed fairly evenly between the new daughter cells. In addition, each daughter cell must have a new nuclear membrane. There is about 700 times as much DNA present, chromosomes number up to 1000 per eukaryotic cell, and there are histones associated with the DNA. The division of a eukaryotic cell into daughter cells occurs in two stages. Its nuclear contents are divided by either mitosis or meiosis and then the cell splits into two by cytokinesis.

Some types of cells normally do not divide but can be stimulated to do so by injury. These include liver and fibroblast cells (a type of proto muscle cell). Muscle cells never divide after tissue differentiation, the development of cells with specialized functions.

During mitosis the cell completes four stages:

- prophase
- metaphase
- anaphase
- telophase

Cytokinesis is the final step. Then the cells enter interphase again and the entire process is repeated (See Figure 5.3).

Prophase is the longest phase. Long, extended DNA-protein fibers (chromatids) coil and condense to become chromosomes. Each chromosome contains two identical chromatids, which are attached at an indentation called the centromere. Meanwhile, the spindle apparatus is built out of protein and it forms in the cytoplasm of the cell. To prepare for the

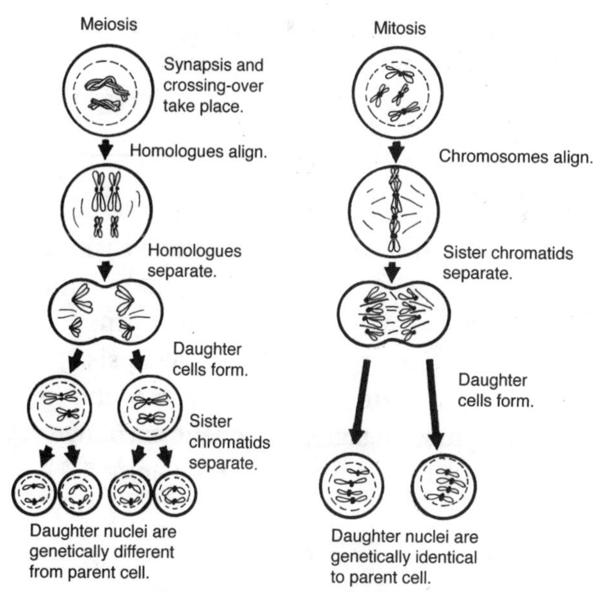

Figure 5.3—Meiosis and Mitosis

split, the chromatids attach to the ends of certain spindle fibers by two brush-like filaments (kinetochores). The nuclear membrane and the nucleolus disassemble.

In metaphase, the chromosomes line up midway between the spindle poles at the equatorial plane. At this point, a human somatic (non-productive) cell has forty-six chromosomes and ninety-two chromatids.

During anaphase, the kinetochores suddenly and synchronously release their attachment. The chromosomes move toward opposite poles. They are separate.

In telophase, the chromosomes reach their respective poles. The chromosomes uncoil, the spindle apparatus disassembles, the nuclear membrane forms around the chromosome cluster, and the nucleolus reassociates.

Mitosis ends when the new daughter cells go their separate ways via cell cleavage, or cytokinesis. Cytokinesis in animal cells is a result of contractions of a ring of microfilaments (actin molecules) just beneath the plasma membrane. As the ring contracts, it pulls the membrane inward, squeezing the double cell in the center until it pinches off into two complete but smaller cells.

- **After cytokinesis is complete, the microfilament ring is broken down and the subunits are used in other microfilaments—an example of nature's habit of recycling.**

Cytokinesis happens differently in plant cells because it isn't so easy to pinch off that tough cell wall. Vesicles from the Golgi apparatus (see page 20) gather at the middle of the mitotic spindle. They contain building materials for a new primary cell wall. Plasma membranes of the vesicles fuse to form a cell plate. Materials inside of the vesicles get sandwiched between two new membranes that elongate along the plane of the cell plate. As cellulose is deposited over both membranes, the cell plate grows at its margins until it fuses with the plasma membrane of the parent cell.

Meiosis

Reproduction may be the most important characteristic that distinguishes the living from the nonliving. Sexually reproducing organisms are made up of two types of cells: germ cells and somatic cells. Germ cells reside in the gonads of the animals—the female ovaries and the male testes. Germ cells undergo meiosis to form the gametes used in sexual reproduction.

- **Somatic cells divide by mitosis.**
- **Germ cells divide by meiosis.**

If meiosis had never appeared, life on earth may well have been limited to bacteria and simple, one-celled eukaryotes! Meiosis increases the genetic variability for the members of the species through sexual reproduction. This is because each human gamete contains twenty-three unpaired chromosomes (making them haploid), rather than the usual forty-six paired chromosomes (which are diploid). When haploid male and female cells join they create a new diploid cell called a zygote with an equal number of chromosomes from each parent. The eventual shuffling of the chromosomes and genes form new genetic combinations, which are distributed to haploid daughter cells.

- **Gametes in females are eggs. In males, they are sperm.**

Unlike mitosis, which maintains the diploid chromosome number, meiosis divides the number of chromosomes in half. During the process, homologues (matched copies for each chromosome) of a diploid mother cell separate into four haploid daughter nuclei. The haploid products ultimately become incorporated into gametes for sexual reproduction. The fusion of two haploid gametes (male and female) restores the diploid number of chromosomes in the zygote, thus maintaining a stable number of chromosomes from one generation to the next.

In human males, diploid cells in the testes undergo meiosis after puberty. In human females, meiosis begins in the fetus. The diploid cells of the ovaries complete development with the onset of puberty—usually one at a time and one per month.

Meiosis takes place in two divisions. Meiosis I is called the reduction division because it separates homologous chromosomes into two different daughter nuclei. The second division separates attached chromatids, producing a total of four haploid nuclei. The names of the phases of meiosis are the same as the names of those in mitosis: prophase, metaphase, anaphase, and telophase. Like mitosis, chromosome replication during an interphase period precedes meiosis I.

Chromosomes replicated in interphase begin to coil. Homologous pairs recognize each other and align themselves along their whole length. That's called synapsis. This forms a structure called a tetrad. During synapsis, loops of DNA extending from each of the chromosomes engage in a genetic exchange process called crossing over (See Figure 5.4).

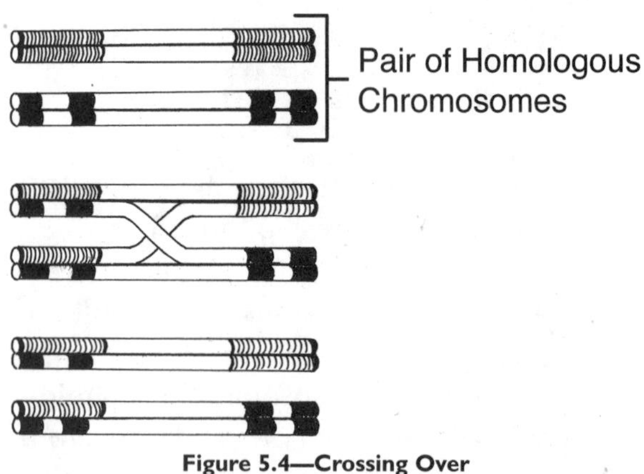

Figure 5.4—Crossing Over

During crossing over, chromosomes exchange portions of chromatids. This is another way that the genes are shuffled, ensuring that each offspring has a random combination of traits from both parents. It is part of the reason why you may look very different from your siblings.

The nuclear membrane and nucleolus disassociate. Spindle fibers become connected to the kinetochores of the chromosomes and the tetrads begin moving toward the center of the cell.

The tetrads become aligned in metaphase I, so that the two homologous chromosomes of each tetrad are on opposite sides of the metaphase plate. It would be very unusual if all the chromosomes from the mother and all those from the father were on opposite sides. They are sorted independently so more genetic variation is possible.

During anaphase I, the spindle fibers shorten and pull the chromosomes of each tetrad toward the opposite poles. The kinetochores do not split (as they do in mitosis). As a result of anaphase I, each daughter cell will have half the number of chromosome sets of its mother cell.

During telophase I, there is some uncoiling of the chromosomes, the nuclear envelope reassociates, and cytokinesis produces two daughter cells. A very brief period called interkinesis separates meiosis I and meiosis II.

When gametes combine during fertilization, the organism produced is different from all others in the species. Together with the appearance of new genetic characteristics by mutation, meiosis is one of the underlying mechanisms that ensures genetic change in a population over time. In other words, meiosis is a key to the formation of new species.

Sister chromatids become separated into different nuclei during meiosis II. In prophase II, the chromosomes shorten and thicken, the nuclear membrane dissociates and the nucleolus disassembles. In metaphase II, chromosomes attach to the spindle and line up at the metaphase plate. During anaphase II, the kinetochores split, pulling the newly independent chromosomes toward opposite poles. In telophase II, the nuclear envelope forms around the four DNA clusters and the nucleolus reappears. Cytokinesis ends the process with four separate haploid cells formed.

CANCER—CELL CYCLE OUT OF CONTROL

Abnormality in cell cycle control can lead to cancer. There are at least three points in the cell cycle that are checkpoints (See Figure 5.1). Before a cell can continue past the G_1 checkpoint, the cell must reach a certain size, and have enough nutrients and growth factors. After that, the cell will replicate its chromosomes. Toward the end of the G_2 phase, there is check to make sure that the cell size is adequate and that DNA replication was complete and successful. Now the cell will enter prophase.

There is a final check at metaphase to make sure that the chromosomes are attached properly to the spindle apparatus. If the chromosomes are not properly attached, they could separate incorrectly and daughter cells would receive the incorrect number of chromosomes. This could be disastrous! The checkpoints at G_1 and G_2 prevent the cell from growing uncontrollably.

Cancer cells exhibit rapid and uncontrolled growth (See Figure 5.5). They grow and divide abnormally. The controls that normally prevent overcrowding in tissues are lost. The plasma membrane and cytoplasm of cancerous cells change. The plasma membrane may become more permeable, resulting in lost proteins. If recognition proteins are altered or lost, the cells have a weakened capacity for adhesion, and cells cannot stay anchored in proper tissues. Different proteins trigger abnormal increases in small blood vessels. In a sense, the cancerous tissue is now providing additional blood to itself! Your circulatory system is bringing more oxygen and nutrients to the growing tumor. The cytoskeleton becomes disorganized and/or shrinks.

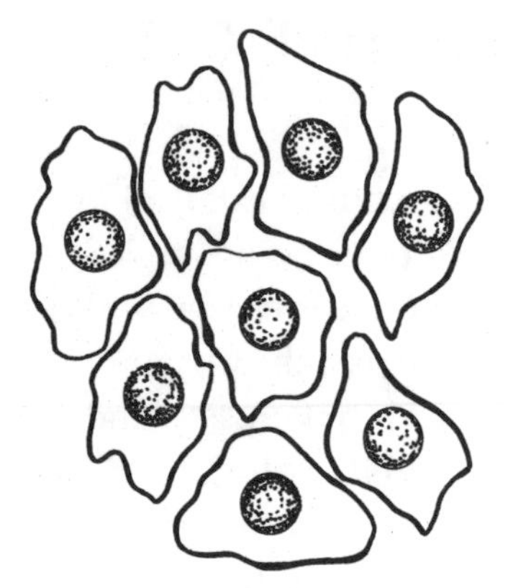

Figure 5.5—Cancer Cells

Causes

Chemical carcinogens (tars, nitrates, and asbestos, for example) can cause local changes in the DNA sequence leading to cancer. Physical carcinogens cause breaks in chromosomes and can lead to translocation. These include ultraviolet light and X-rays. Exposure to the sun's UV rays is a leading cause of skin cancer. Cancer-causing (oncogenic) viruses introduce foreign DNA into cells. The viral DNA gets incorporated into the host cell DNA and the viral proteins produced may influence the host's cell division. Oncogenes (any gene having the potential to induce a cancerous transformation) have been implicated in some cancers. Chromosome #17 has a gene which codes for a protein—p53. If a cell is deficient in p53 or it is altered, it does not enter G_0. Alterations to the gene and, therefore, to this protein, yield the most frequently encountered genetic event in human malignancy. Another protein, p16, impedes cell division. If there is no p16 or a malfunctioning p16 is produced, cell division is not restrained. This has been implicated in malignant melanoma.

Treatment

There are many chemotherapy treatments available to treat and cure cancer. Let us look at four of these, and see their impact on the cell cycle.

Most chemotherapy treatments work on rapidly growing cells. Hair follicles and the cells lining the mouth and esophagus, bone marrow, and skin cells normally divide rapidly. Chemotherapeutic agents affect these cells. Methotrexate inhibits DNA synthesis and has been used to treat leukemia. Vinblastin blocks microtubule assembly and is used in the treatment of leukemia and Hodgkin's disease. 5-Fluorouracil inhibits DNA synthesis and is used for breast, stomach, and colon cancers. Another drug, cyclophosphamide, alkylates DNA bases, preventing entry into the S phase. This has been used successfully in the treatment of lymphomas, and breast and lung cancer.

Cancer is not a single disease but a collection. What works for one may not have an effect on another. Nonetheless, researchers are making tremendous progress on many fronts. Perhaps, someday, cancer will be at worst controlled, and at best eradicated.

SUMMARY

- The cell cycle is composed of an interphase, during which the cell is metabolically active, growing larger, and replicating its chromosomes. Once the cell has reached a certain size and the DNA has doubled, the cell will undergo division.
- Prokaryotes divide by binary fission. The two daughter cells that are produced are virtual clones of the each other—no genetic variability—unless a mutation has occurred.
- Eukaryotic cells undergo mitosis or meiosis. All somatic cells divide by mitosis. DNA replication occurs during interphase before nuclear division begins. There is one nuclear division that has four stages: prophase, metaphase, anaphase, and telophase. Mitosis ends with cytokinesis.
- Meiosis is a longer process than mitosis and only occurs in the germ cells of sexually reproducing organisms. DNA replication occurs once, during the interphase before meiosis I. There are two divisions and each includes four stages: prophase, metaphase, anaphase, and telophase. Meiosis forms four, haploid cells, which contain half as many chromosomes as the parent cell.
- Cancer is the result of a cell whose cell cycle is out of control. This is generally due to the absence or alteration of a protein that is necessary to prevent cell division. The cells divide rapidly and may spread to other parts of the body. Chemical and physical carcinogens may cause cancer. There is a genetic disposition for some types of cancer.
- The mode of action of the chemicals used in the treatment of cancer target certain sites of the cell cycle. Many prevent the replication of the chromosomes.

CHAPTER 6

GENETICS

KEY TERMS

homozygous	heterozygous	genotype
alleles	phenotype	dominant

Do you have your mother's eyes or your grandfather's chin? Biological inheritance is the key to differences among humans and, of course, species. In the early nineteenth century, people thought that traits from parents simply "blended" in the child. If the blended theory was correct, however, once traits were blended, they should not have appeared again. Populations should have reached a uniform appearance after many generations. All you need to do to disprove that is look around in a crowded mall. *Variation* in populations is what we see—not uniformity. Many of the answers about diversity came from studies with pea plants.

GREGOR MENDEL

Gregor Mendel, an Austrian monk and high school teacher, was born in 1822 and grew up on a farm. Very bright, he went to the University of Vienna and studied science and mathematics. He was particularly influenced by fellow Austrians, Christian Doppler, a physicist, and Franz Unger, a botanist. Doppler taught him to apply a quantitative experimental approach to the study of natural phenomena. Unger sparked his interest in the causes of inheritable variation in plants.

- **Christian Doppler discovered the Doppler effect in 1845. It is the change in pitch of sound as it moves toward you and away from you.**

Experiments with Peas

Why study peas? Mendel was in charge of the monastery garden. He had studied biology at the university. From his studies, Mendel had gained an understanding of the sexual mechanisms in the pea plant. He knew that each plant was monoecious, that is, the pollen from a pea plant can fertilize the female eggs of the same plant. Self-fertilization provides strict control over mating. Seeds produced by self-fertilization inherit only traits present in the plant that bore them. In addition, Mendel had learned how to prevent self-fertilization and performed many cross-pollination experiments.

- **Certain peas have many easily distinguishable traits, cost little, and reproduce quickly. All of these attributes made the pea plant a good organism to study.**

Keep in mind that the process of meiosis and the concept of genes were unknown in Mendel's time. Perhaps the single most important decision made by Mendel was to study a few isolated characteristics (traits) of the pea plant. This made his job of measuring the effects of heredity much easier. Mendel examined seven traits in the pea plant: flower color, flower position, seed color, seed shape, pod color, pod shape, and stem length. Each characteristic only occurs in two contrasting forms—dominant and recessive.

PEA TRAITS STUDIED BY MENDEL

Trait	Dominant	Recessive
Flower Color	Purple	White
Flower Position	Axial	Terminal
Seed Color	Yellow	Green
Seed shape	Round	Wrinkled
Pod Color	Green	Yellow
Pod Shape	Inflated	Constricted
Stem Length	Tall	Dwarf

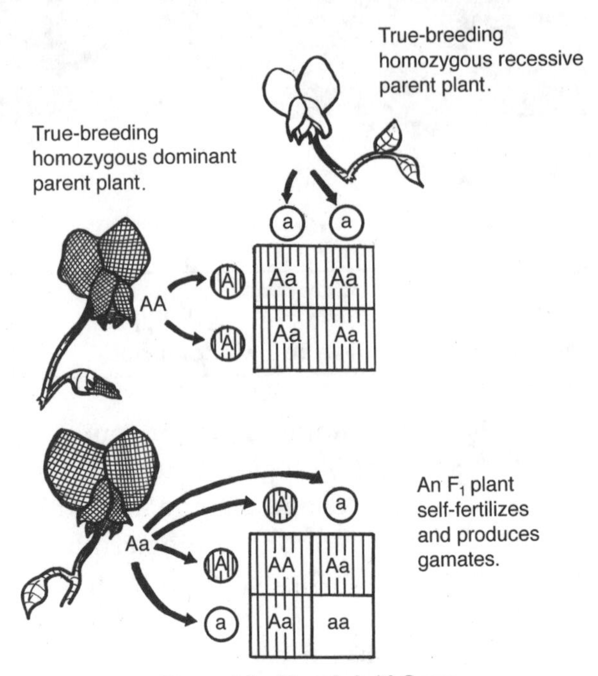

Figure 6.1—Monohybrid Cross

Monohybrid Crosses

Mendel allowed thousands of pea plants to self-fertilize. After obtaining true-breeding (see page 43) plants of all seven traits, Mendel was ready to further explore heredity. Mendel tracked the traits over three generations: (P) the parent generation; (F_1) the first generation of offspring; and (F_2) the second generation of offspring. For one set of experiments, he crossed true-breeding purple-flowered plants with true-breeding white-flowered plants. Mendel removed stamens from a purple flower. Using a small artist's paintbrush, he transferred pollen from the stamens of a white flower to the carpels of a purple flower. The pollinated carpel matured into a pod. He planted the seeds from the pod. All the offspring of the F_1 generation had purple flowers. Figure 6.1 illustrates this genetic cross.

Mendel took this a step further. He allowed F_1 heterozygous plants (having two different genes for the same characteristic) to self-fertilize and produce gametes. Most of the offspring of the F_2 produced purple flowers, but about twenty-five percent produced white flowers. The plants' phenotypes, or genetically determined appearance, was a 3:1 ratio of purple flowers. There wasn't a light purple flower among the offspring. This blew away the blending theory of heredity!

Dihybrid Crosses

Mendel used dihybrid crosses to explain how two pairs of genes are distributed in the gametes. He crossed true-breeding tall pea plants with purple flowers and true-breeding dwarf pea plants with white flowers. All of the offspring of the F_1 generation had the same phenotype—tall pea plants with purple flowers (See Figure 6.2). When he allowed the hybrid pea plants to self-fertilize, he discovered a phenotypic relationship of 9:3:3:1 among the plants in the F_2 generation:

9 tall with purple flowers

3 dwarf with purple flowers

3 tall with white flowers

1 dwarf with white flowers

Theory of Segregation

Mendel believed that plants inherit information about a trait, receiving one unit from each parent. To find this out, he performed many monohybrid crosses. He crossed a purple-flowered pea plant from the F_1 generation with a true-breeding, white-flowered plant. If the hybrid were homozygous (having two identical genes for one characteristic) for purple flowers, all of the offspring from the testcross would be purple. Offspring with white flowers meant the F_2 individual must have been heterozygous for the purple-flower trait. Mendel concluded that the "units" of heredity from the parents must segregate separately in fertilization.

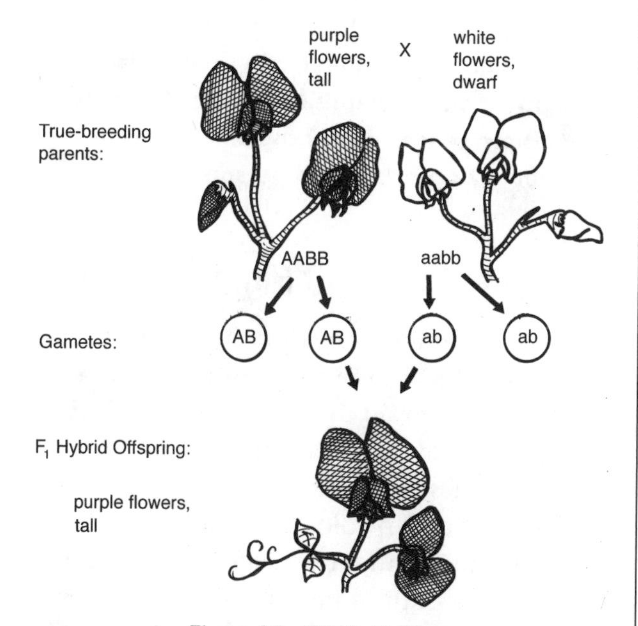

Figure 6.2—Dihybrid Cross

- **Thomas Hunt Morgan coined the term "genes" in 1910. He experimented on heredity in fruit flies.**

Mendel developed a four-part hypothesis. First, he suggested that alternative forms of inherited units are responsible for variations in traits. Second, for each trait, an individual inherits two alleles, one from each parent.

GENETIC TERMS TO KNOW

Term	Definition
allele	One of a number of different forms of the same gene for a specific trait.
dihybrid cross	Tracking the inheritance of two traits between two individuals.
dominant trait	The expressed trait in a heterozygous individual.
gene	Unit of heredity consisting of a particular sequence of nucleic acid that codes for a certain protein.
genotype	Genetic makeup of an individual.
heterozygous	Both alleles of a gene are different.
homozygous	Both alleles of a gene are identical.
monohybrid cross	Tracking the inheritance pattern of a single trait between two individuals.
phenotype	Expressed traits of an individual.
recessive trait	The trait is not expressed in the heterozygous individual.
test cross	Mate an individual with an unknown genotype for a specific trait with an individual who is homozygous recessive for the same trait.
true-breeding	Sexually reproduced organisms with inherited trait(s) identical to parents.

Third, if the alleles differ, one is expressed (dominant); the other is masked (recessive). Finally, the two alleles segregate during gamete production. This is astounding since he had no idea what a gene was, let alone what the process of meiosis was! Mendel's Theory of Segregation still stands. We now know that diploid cells have two sets of homologous chromosomes. The coding sections of these chromosomes are the genes. The two genes of each pair are separated from each during meiosis (See Chapter 5) and, therefore, end up in different gametes.

Theory of Independent Assortment

As a result of the dihybrid crosses, Mendel thought that the two units for the first trait he was tracking had been distributed into gametes independently of the other two units for the other trait. We now know that this theory can be explained by the process of meiosis. By the end of meiosis, each pair of alleles segregates independently into gametes.

BEYOND MENDELIAN GENETICS

Incomplete Dominance

In incomplete dominance, the dominant phenotype is not fully expressed in the heterozygote. The phenotype ends up somewhere between the dominant and the recessive. Snapdragons provide a good example of this genetic concept. If you cross a homozygous red flower with a homozygous white flower, all of the F_1 offspring are pink. Sounds a lot like that "blended" theory again, but it isn't. If you cross two pink snapdragons, the F_2 offspring yield the following phenotypes: 1 red: 2 pink: 1 white. This is also proof that blending of traits does not occur, or the red flower would not have been produced. Biochemical studies on the snapdragons have shown that the red color is due to the production of a red pigment. There is only half as much pigment in the pink flowers, and there is no pigment in the white flowers. So it isn't that the red and white blend to make pink. Half the amount of pigment produces a pink flower.

Codominance

Sometimes both alleles are expressed. Have you ever wondered how letters are assigned to blood type? There are multiple alleles for human blood types: A, B, and O. In codominance, both alleles are expressed in the heterozygotes. So, if you inherit A from one parent and B from the other parent, you will have a blood type of AB. A and B (antigens, or proteins that elicit an immune response) are genetically determined polysaccharides (long chain of simple sugars) found on the surface of the red blood cells.

- **More than one gene is responsible for skin color in humans. Three genes affect skin color equally. Melanin is the brown-black pigment that is deposited in the skin cells. Multiple genes control the differences in the amount of melanin in a person's skin.**

Multiple Effects of Single Genes

In some cases, the expression of a pair of alleles is influenced by the genotype at another location on the chromosomes. This may be illustrated by human pigmentation. Separate genes are responsible for hair color, eye color, and skin color. An individual affected by albinism will have a lack of color in the hair, eye, and skin. An enzyme is missing in this individual that prevents the formation of the pigment melanin. Albinism is an example of epistasis—a pair of recessive alleles at one site masks the effect of genes at another site.

Another example of epistasis occurs in mice. Black mice are homozygous dominant or heterozygous for color. Brown mice are homozygous recessive for color. In order for the black or brown color to be expressed, the mouse must also have one allele at a different site to allow the black or brown pigment to be deposited in the fur. If that is not present, the mice will be white.

Oncogenes

Any gene that has the potential to produce a cancerous transformation is referred to as an oncogene. These genes were first identified in certain RNA viruses. Oncogenes are altered forms of normal genes. Transformation of any cell may start with mutations in the oncogene itself. This may happen if there is a mutation in the DNA or the insertion of viral DNA into the cell's DNA (in other words, viral infection). Ultraviolet radiation, X-rays, gamma rays, asbestos, and components of cigarette smoke can also cause changes in the DNA. Cancer is a multistep process and probably involves greater than one oncogene. For some colon cancers, three oncogenes have been discovered thus far. The product of these oncogenes allows cells to divide in an uncontrolled fashion.

The Human Genome Project

The Human Genome Project, begun in the early 1990s, was an international collaborative effort of many scientists to map and sequence all of the genes of the human genome. It is largely finished, well ahead of its original goal date of 2005. The DNA sequencing of the genes was the most arduous part of this project.

The potential benefits of having a complete map of our genes are enormous. It will give insight into embryonic development and evolution. The knowledge of the gene sequence may have a major impact on the diagnosis, treatment, and most importantly, the prevention of some disease.

HUMAN GENETICS

Autosomes Versus Sex Chromosomes

Males and females have forty-four autosomes and two sex chromosomes. An autosome is a nonsex chromosome. There may only be two of them, but the sex chromosomes are critically important. They determine the gender of the individual. Females have XX chromosomes and the males have XY chromosomes. The Y chromosome is much smaller than the X. Therein lies the reason behind sex-linked disorders. A trait carried on the X chromosome in the male, frequently will not have an allele on the Y to mask it, and, therefore, it will be expressed. Most of the alleles on the Y chromosome govern sexual characteristics in the male, such as development of male genitalia, as well as secondary characteristics, such as enlarged larynx and hair distribution. The Y chromosome does not carry alleles that might counteract (mask) an otherwise recessive genetic disposition on the X chromosome—say, color-blindness. Such a disorder isn't as likely to show up in female offspring, but the gene for it is on the X chromosome; therefore the mother is the carrier.

During fertilization, since the female can only donate an X chromosome, the gender of the offspring is solely determined by the male, who can donate either an X or a Y chromosome.

Patterns of Inheritance

Genetic disorders involve defective alleles, which usually code for either a malfunctioning protein or no protein at all. There are three types of disorders linked to patterns of inheritance: autosomal recessive, autosomal dominant, and sex-linked disorders. Let's look at some examples of each.

Autosomal Recessive Disorders

Individuals who are carriers (heterozygotes) for autosomal recessive disorders have a twenty-five percent chance of having children with the genetic disease. Heterozygotes can be phenotypically normal. Albinism is a non-lethal, autosomal disorder. The outer surface of the individual appears white. The pink color of his eyes is due to the absence of melanin from a tissue layer in the eyeball, which allows red light to be reflected from blood vessels in the eye. The enzyme, tyrosinase, necessary in the formation of melanin, is missing. Individuals with albinism are particularly sensitive to the UV rays of the sun.

Sickle-cell anemia is the most commonly inherited disease among African Americans. Approximately one in five hundred African Americans born in the United States has this disorder. It is caused by a single amino acid substitution in hemoglobin, resulting in the production of a malformed protein. The ability of a gene to have multiple phenotypic effects is called pleiotropy and can be illustrated by the sickle-cell gene.

The abnormal hemoglobin crystallizes when the oxygen content of blood is low, causing red blood cells to become sickle-shaped. The red blood cells break down and clump, clogging capillaries. Extra work is placed on the spleen to rid the body of these cells, ultimately leading to damage of the spleen. A cascade of events occurs. The breakdown of the red blood cells causes physical weakness, anemia, and heart failure. Clogged capillaries can lead to pain and fever, heart failure, and damage to the brain and other organs. Treatment includes many blood transfusions.

SICKLE-CELL ANEMIA

1. Abnormal hemoglobin crystallizes—Red blood cells change shape
2. Red blood cells break down = weakness, anemia, heart failure
3. Cell clumping, blood vessels clog = pain, fever, brain damage, organ damage
4. Cells accumulate in spleen = spleen damage

Long Term Damage

paralysis infections rheumatism kidney failure

Since those with sickle-cell anemia die at a young age, what is the selective advantage of having this gene? Those who are carriers of this allele are less susceptible to malaria. This is particularly important in Africa. A high incidence of the sickle-cell gene is found in sub-Saharan Africa, where malaria is found. Therefore, individuals who are heterozygous for sickle-cell anemia will not die of this genetic disorder, nor will they succumb to malaria.

The membrane protein that controls chloride traffic across the plasma membrane is defective or absent in cystic fibrosis. Symptoms include thickened mucus of the liver, intestinal tract and lungs. This leads to bacterial infections. There is no cure, but inhaling the enzyme deoxyribonuclease helps to degrade the DNA that contributes to the viscosity of the mucus.

The DNA is derived from disintegrating inflammatory cells that move into the respiratory tract. Prior to 1992 there was no treatment. The only thing that could be done was to pound on the person's back to dislodge some of the mucus. What is the selective advantage of the heterozygous state of this gene? Only more research and time will tell us that answer.

- **Cystic fibrosis is the most common lethal genetic disease in the United States—one in eighteen hundred Caucasians are affected.**

Tay-Sachs disease is more prevalent among eastern European Jewish descendants, French Canadians, and those of Cajun ancestry. It usually causes death by age five. While normal-looking at birth, children develop symptoms by six months of age. Their development slows down and eventually they lose motor skills and mental functions. They become blind, deaf, mentally retarded, paralyzed, and ultimately non-responsive to their environment. Those with Tay-Sachs are lacking the enzyme hexosaminidase A. They cannot metabolize gangliosides, a type of lipids that accumulate in the brain. Could the heterozygous condition possibly offer an advantage? At present, we do not know.

Autosomal Dominant Disorders

Lethal dominant alleles are much rarer than lethal recessive genes. In most cases, the homozygous dominant individual is spontaneously aborted. Achondroplasia, a type of dwarfism, is an example of such a manifestation. The individual has a normal size head and torso, but the arms and legs are much shorter. Those with two abnormal genes for this condition generally die before birth.

Huntington's disease is a late-acting lethal autosomal dominant disorder. The defected gene has been identified—near the tip of chromosome #4. Symptoms of the progressive deterioration of the nervous system begin at thirty-five or forty years of age. The individual usually dies within five years of the onset of symptoms.

Sex-Linked Disorders

The X chromosome is much larger than the Y chromosome in humans. Most of the genes on the Y chromosome encode traits found only in males. Most X-linked (sex-linked) genes have no homologous location on the Y chromosome, and, therefore, will be expressed. Colorblindness, hemophilia, and muscular dystrophy are examples of sex-linked disorders.

- The most common type of **colorblindness** is red-green colorblindness. There is a malfunction of the light-sensitive cells of the eye. People afflicted with this condition have difficulty seeing the lighter shades of red and green.
- The individual with **hemophilia** is missing a protein that is necessary in the clotting of blood. A small cut could be lethal to such a person because they could bleed to death. Treatment includes the injection of the clotting factor from donated blood.
- **Muscular dystrophy** results in the progressive wasting away of skeletal muscle. It does not affect smooth muscle tissue. Symptoms show up in early childhood, and by the age of twelve, the child is wheelchair bound. Death occurs by the early twenties due to respiratory failure.

A sex-influenced trait is one that is caused by a gene whose expression differs between males and females. Baldness is a sex-influenced trait. It is controlled by two alleles. The difference lies in the heterozygous condition—for example, males are bald and females will have hair. The female will become bald only in the homozygous recessive condition of the same gene.

Chromosomal Mutations

Mutations occur due to alterations in chromosome structure or number (See Figure 6.3). These can be spontaneous or environmentally caused by such agents as ultraviolet light, radiation, chemicals and viruses.

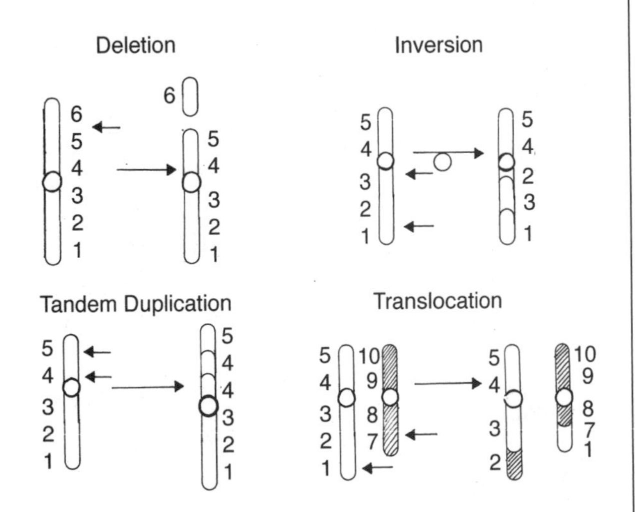

Figure 6.3—Chromosomal Mutations

> A deletion in a chromosome causes the loss of a segment of DNA. This is implicated in the genetic condition called *cri du chat* syndrome. The cry of this infant sounds like a cat mewing—hence the name. The baby has a small head with unusual facial features, is mentally retarded and dies in infancy.

An inversion does not change the amount of DNA in the cells; instead, the order of the bases is reversed for a specific segment of the DNA. If a segment of the DNA duplicates, there is additional DNA in the cells.

Translocation involves an exchange of chromosome parts. The most common type of translocation is seen when nonhomologous chromosomes exchange fragments. In some leukemias, chromosomes #9 and #22 exchange parts. In some individuals with Down's syndrome, part or all of chromosome #21 is attached to a different chromosome.

Change in Chromosome Number

The number of chromosomes in humans is usually forty-six. If the chromosomes or sister chromatids fail to separate correctly (nondisjunction) during cell division, the result could be an incorrect chromosome number. If the homologous chromosomes fail to separate in meiosis and meiosis II is normal, two of the cells formed will have an extra chromosome and two will be short one chromosome. If there is a normal meiosis I followed by nondisjunction (failure of the sister chromatids) in meiosis II, two of the cells will have the normal complement of chromosomes, one will have an extra chromosome, and one will be short a chromosome.

Approximately one in seven hundred children in the United States has Down's syndrome. This is the result of nondisjunction of chromosome #21, giving the individual a chromosome count of forty-seven. These individuals share some physical characteristics including almond-shaped eyes and a larger than normal tongue. They have varying degrees of mental retardation, are short in stature, are sexually underdeveloped (usually sterile), and have a shorter life span. They are more prone to having respiratory infections, heart defects, leukemia, and Alzheimer's disease.

Nondisjunction of sex hormones can also occur. In Klinefelter's syndrome, the individual has forty-seven chromosomes (XXY). The individual with Klinefelter's syndrome has normal intelligence. The male sex organs, the testes, are small and the individual is sterile.

Turner's syndrome is the only viable monosomy in humans; that is, they have only one of a pair of chromosomes. These individuals have forty-five chromosomes. They have only one sex chromosome. Since it is the X chromosome, phenotypically they are females. Their sex organs do not develop and, therefore, they are sterile. No secondary sexual characteristics develop (such as facial hair in males). This individual is less than average height.

SUMMARY

- Gregor Mendel studied the reproduction of pea plants and followed the passing of genetic traits for three generations in these organisms. His genetic experiments and mathematical analyses led him to two theories: the theory of segregation and the theory of independent assortment. These are amazing conclusions, since meiosis was unknown in the late 1800s.
- Future genetics studies have determined other types of inheritance beyond the classical inheritance patterns of Mendel.
- Oncogenes have been implicated in the cause of several human cancers. The Human Genome Project has mapped the genes of humans. This breakthrough work could greatly help in the diagnosis, treatment and prevention of human disorders.
- There are many types of genetic defects. Some are autosomal recessive (sickle-cell anemia, for example), fewer are autosomal dominant (e.g., Huntington's disease).
- Sex-linked disorders, such as hemophilia, are found in males almost exclusively. Nondisjunction of chromosomes leads to an abnormal number of chromosomes. An extra chromosome #21 is the genetic basis for Down's syndrome. Nondisjunction of the sex chromosome can also occur as in Klinefelter's syndrome (XXY) and Turner's syndrome (a single X chromosome).

DNA REPLICATION, TRANSCRIPTION, AND TRANSLATION

KEY TERMS

helix	polymerase	deoxyribonucleic acid
codon	genome	ribonucleic acid
peptide		

If you have ever copied a recipe incorrectly, you know that the results can be unusable. Nature's recipes for life, or hereditary instructions, are no different. Scientists of the 1940's had one of the world's greatest mysteries on their hands. Biologists knew that genes were reshuffled during meiosis and passed on from one generation to the next. But what carried the hereditary instructions—DNA or protein? The nature of the genetic information eluded them. Let's take a look at how the mystery was solved.

DISCOVERY OF DNA AS THE GENETIC MATERIAL

Frederich Miescher first purified DNA in 1869. He isolated the DNA from two sources: salmon sperm cells which have very large nuclei, and pus cells obtained from discarded surgical bandages.

Phoebus Levene identified the sugar ribose in some nucleic acids in the early 1900's. In 1929, he discovered the sugar deoxyribose in DNA. This was an important step, since this was a major chemical distinction between RNA and DNA. He further determined that a nucleotide was composed of a five-carbon sugar, a phosphate portion, and one nitrogenous base (See Figure 7.1). We had to wait for Edwin Chargaff to come up with the correct proportion of bases in the late '40s. If the four bases were found in equal amounts, the coding allowed by DNA would be limited.

Thymine (T)

Cytosine (C)

Adenine (A)

Guanine (G)

Figure 7.1—Nitrogenous Bases of DNA

- **Robert Feulgen, in 1923, developed a procedure that specifically stained DNA in cells undergoing mitosis.**

Experimental evidence that DNA is the genetic material begins with Fred Griffith in 1928. He was working with two strains of *Streptococcus pneumoniae*—a round-shaped bacterium. One strain had a polysaccharide capsule (S-strain—smooth) and was pathogenic (creates disease) and the other lacked an outer capsule (R-strain—rough) and was nonvirulent. When he injected mice with the S-strain, the mice died from pneumonia. He was able to recover living, smooth virulent bacteria from a dead mouse. The capsule protected the bacterium from the mouse's immune system. When he injected mice with the R-strain, the mice lived. The mouse's immune system was able to destroy the bacterium. Next, Griffith heat-killed the S-strain and injected the mice. The mice lived. Then, he heat-killed the S-strain and added living R-strain. Both were injected into mice, and the mice died of pneumonia. Living, smooth virulent bacteria were recovered from the dead mice. Griffith's experiments demonstrated that a biochemical agent in the virulent strain of *S. pneumoniae* could make a nonvirulent strain deadly. Griffith did not know the transforming agent, but hinted that it was not protein because he knew that heat denatured protein.

Oswald Avery, Colin MacLeod, and Maclyn McCarthy repeated Griffith's experiments. They added a soluble extract of the encapsulated cells to noncapsulated cells in a growing medium. Transformation of the bacteria occurred! Avery, MacLeod, and McCarthy spent ten years purifying this substance. By adding either a protease or an enzyme that disrupts DNA to Griffith's experiment, they demonstrated that DNA—and not the protein—transmits the ability to kill. When the protease was added to the S-strain and injected into mice, the mice died. When deoxyribonuclease was added to the S-strain and injected into the mice, they lived. Avery, MacLeod, and McCarthy discovered that DNA was the transforming agent in 1944.

The work of Avery, MacLeod, and McCarthy was overlooked. Why? These scientists were bacteriologists. Many doubted that studies with bacteria could be relevant to higher organisms. Perhaps transformation was a fluke, they thought. Finally, some believed that there could be protein contaminants. It is the nature of science to look for ways to discount or disprove new ideas.

The scientific community needed further proof that DNA was the hereditary material. Alfred Hershey and Martha Chase provided this proof in 1952 with the famous experiment that demonstrated that viral DNA could program cells. They used either radioactive sulfur (found in proteins but not DNA) or phosphorus (found in DNA but not in proteins) to label bacteriophages. Hershey and Chase were able to prove, beyond any doubt, that DNA is the genetic material.

The elegant experiment by Hershey and Chase was really very simple. It was based on the fact that some viruses are made up of just DNA with a protein coat. When these tiny viruses infect a much larger bacterium, they take over the chemical reactions that synthesize DNA and protein. Hershey and Chase wanted to know which part of the virus was responsible for this takeover. To find out, they prepared two special groups of virus. In one, the viruses were grown in an environment where all of the sulfur was radioactive. This labeled all the viral proteins, but not the viral DNA because there is no sulfur in DNA. A second group of viruses was grown with radioactive phosphorus. This labeled all of the viral DNA, but not the proteins because the phosphorus is very abundant in DNA but not protein. Then, they used these two batches of viruses to infect two different populations of *E. coli* bacteria.

When the infections were complete, Hershey and Chase stopped the experiments by breaking up all the bacteria and viruses in a blender. They then centrifuged these homogenates and got pellets of infected cells at the bottom of the centrifuge tubes. The rest of the tubes were filled with a clear cell-free fluid, the supernatants. Interestingly, the radioactive sulfur from the viruses was all in the cell-free supernatants, which demonstrated that the viral proteins were not incorporated into the in-fected *E. coli* bacteria. On the other hand, the radioactive phosphorus from the viruses was all found in the infected bacteria. This result clearly demonstrated for the first time that the part of the virus that entered the bacterium and caused mass production of more viruses is the DNA.

DNA—THE DOUBLE HELIX

One by one, the clues were falling into place. Evidence that DNA is the genetic material of eukaryotic cells was being collected. You've already seen that the DNA of eukaryotic cells doubles its DNA content prior to mitosis. This doubled DNA is equally divided between the daughter cells. The diploid cells of a species have twice as much DNA as that of the species' haploid cells. Edwin Chargaff provided experimental proof in 1947. His work confirmed that the DNA composition varies from species to species. He found that the amount and ratios of the nitrogenous bases varies among the species. In addition, he found a regularity of base ratios within each species: the amount of adenine (A) equals the amount of thymine (T), and the amount of guanine (G) equals the amount of cytosine (C).

In the early 1950's, Maurice Wilkins and Rosalind Franklin bombarded DNA with X-rays, employing a technique called X-ray diffraction. A regularly repeating structure of nucleotides was apparent from these X-rays.

Finally, in 1953, James Watson and Francis Crick put the pieces of the DNA puzzle together and proposed the structure of DNA was a double helix. It was the only molecular shape that worked with the data. Looking at the X-ray crystallography of Wilkins and Franklin, Watson and Crick concluded that DNA is a double helix with a uniform width of 2.0 nm (nanometer—one thousand millionth of a meter). The purines (A and G) and pyrimidines (T and C) are stacked 0.34 nm apart, and there is a full turn every 3.4 nm along its length. There are ten layers of nitrogenous bases in pairs in each turn of the helix. In all these respects, the Watson-Crick model of DNA structure is consistent with the known data. The pattern of base pairing (A=T and G=C) is consistent with the known composition of DNA. The two strands of the double helix are held together by hydrogen bonds between the bases. The mystery was solved!

- **The 1962 Nobel Prize in Medicine was awarded to Watson, Crick, and Wilkins for discovering the structure of DNA.**

DNA REPLICATION

Given the critical importance of DNA as the basic instruction set for life, it's clear that DNA must be copied exactly from cell to cell. To make sure that this process of DNA replication takes place without errors, there are many controls and regulations. Of course, this means that many different enzymes and proteins are involved, which complicates the process, but it follows a highly organized series of three main steps: initiation, elongation, and termination.

A group of specific protein "switches" *initiates* the process of replication by first untwisting the ladder structure of DNA. As the DNA helix untwists, the DNA divides into two halves, which exposes the individual rungs of the ladder that contain the four nucleotide bases (adenine, thymine, cytosine, and guanine) that code all our genetic information. As these nucleotide bases become exposed, the enzyme DNA polymerase matches the exposed bases to their complementary bases. Adenine always pairs with thymine and cytosine always pairs with guanine. Using this code, DNA polymerase always adds the correct nucleotide while building the new DNA strands. This process of *elongation* is fast enough to add about 500 nucleotides per second in single-celled organisms, and copies the entire human genome (about 6,000,000,000 nucleotides!) in just a few hours. And this all happens with only about one mistake in every 1,000,000,000 nucleotide pairings (See Figure 7.2).

The structure of DNA is highly specialized to be "read" in only one direction, from the starting end, which we call the 3′ end, (the little symbol stands for prime) to the finishing end, which we call the 5′ end. When two strands of DNA match up, the 3′ end of one strand matches up with the 5′ end of the other strand. This structure is very stable, but it means that the two strands run in opposite directions. An important result of this difference in direction is that the molecule DNA polymerase also runs in different directions along the two strands. DNA polymerase "reads" the template strands only in the 3′ to 5′ direction, which means that the two strands elongate in different directions. Because one strand of the DNA always runs from the 5′ to 3′ direction, this part of the DNA is elongated in short pieces called Okazaki fragments (named for Reiji Okazaki, who described them in 1968), that are later joined together by an enzyme called DNA ligase. When the whole process of elongation

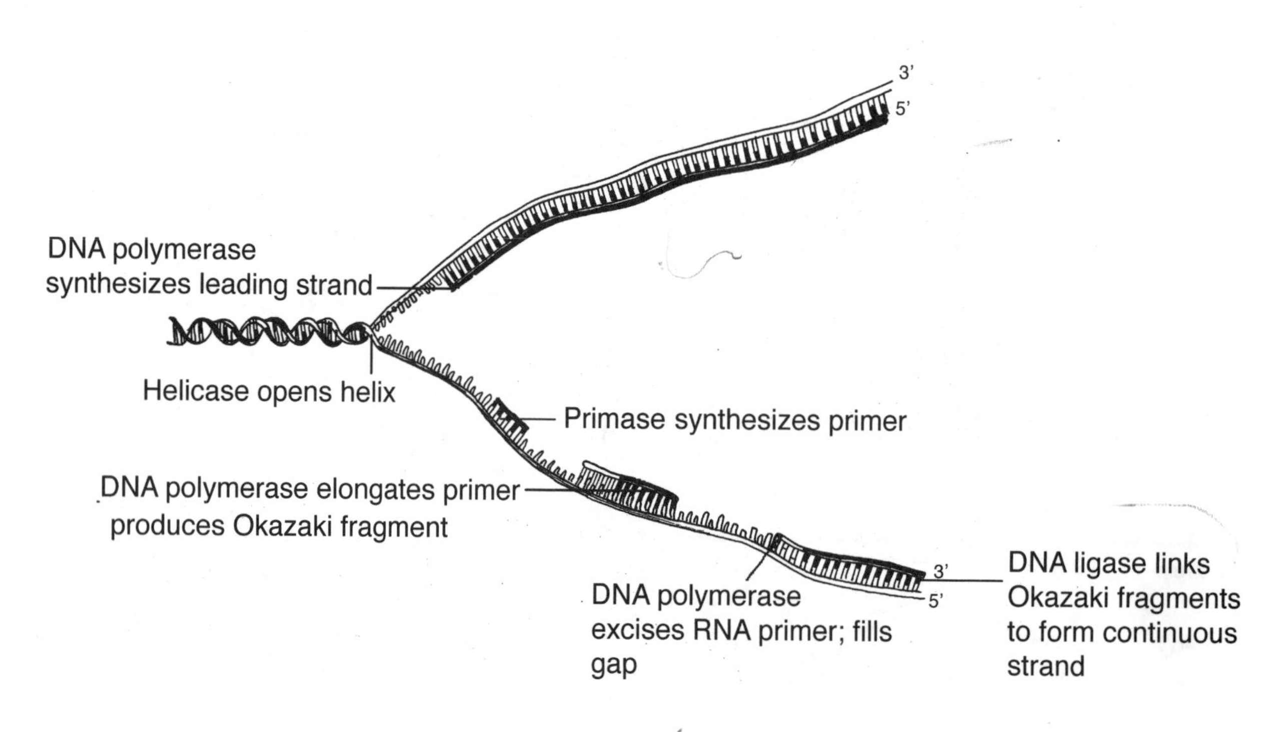

Figure 7.2—DNA Replication

has been completed for both original strands of DNA, the process is shut down and the new strands of DNA are packaged in a process called *termination*. The net result is that all the genetic information carried by DNA has been duplicated so that new cells will have their own exact copy of all the genetic instructions needed for life.

DNA TRANSCRIPTION

The information stored in DNA is a set of instructions for how to make each protein in the body (much like a computer program listing). These instructions are stored as a sequence of nucleotides we call genes. Genes can be short or long, can exist as multiple copies in different places in the DNA, and are copied in their entirety during the process of DNA transcription (See Figure 7.3). For the most part, the goal of this process is just to make a copy of the information in a gene and then send this copy to specialized structures in the cytoplasm where it can be used to make new protein. In eukaryotic cells that have a nucleus, this readout of DNA information takes place in the nucleus.

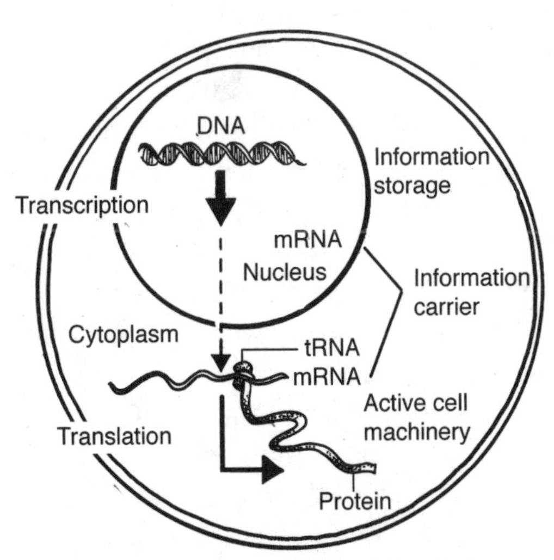

Figure 7.3—Transcription and Translation

The process of DNA transcription follows the same three basic steps used in DNA replication: initiation, elongation, and termination. Unlike DNA replication however, the end product of transcription is a single-stranded molecule of ribonucleic acid. RNA is similar to DNA in that it too contains adenine, guanine, and cytosine. RNA, however, does not contain thymine, and instead contains the nucleotide uracil.

> There are three main classes of RNA. Transcription of most genes produces mRNA (messenger RNA), which carries the protein-building codes. Transcription of other genes produces rRNA (ribosomal RNA), which is the major component of ribosomes, the specialized protein-building factories located in the cytoplasm. Transcription of still other genes produces tRNA (transfer RNA), which is also involved in protein synthesis (see DNA Translation).

During initiation, a selected gene is targeted and the DNA that makes up this gene is exposed through a series of enzymatic reactions so that the enzyme RNA polymerase can bind to the starting point for the readout of the gene, called the initiation site. The RNA polymerase then separates or unzips the two DNA strands, and simultaneously synthesizes a single strand of RNA via the process of elongation.

- **The letter R in RNA stands for the sugar ribose, which is unique to RNA (in comparison, the D in DNA stands for deoxyribose).**

The end of the gene is usually indicated by a special terminator sequence. When this sequence is reached, RNA polymerase activity stops, the halves of the DNA rejoin, and the RNA is released. Often, this RNA is modified before it is sent out to the protein synthesizing machinery. For example, extra or unused instructions are cut out, as are sequences added that simply stabilize the RNA during its synthesis. Other modifications can be made that affect how the RNA moves through the cell, how long it lasts, etc. Indeed, many of these modifications are still a mystery because we don't really understand why they are important. When all of these modifications to the RNA are complete, the RNA is ready for the next step in the process: DNA Translation.

DNA TRANSLATION AND THE GENETIC CODE

DNA translation is the process that converts the information stored in the mRNA synthesized during DNA transcription, into protein (Figure 7.3—Transcription and Translation). This process takes place on the surface of ribosomes, specialized structures made in the nucleolus and then exported to the cytoplasm. Ribosomes are made of two parts or subunits of two different sizes, big and small. Functioning ribosomes include one subunit of each size. Where the ribosomes are located is also important. For the synthesis of proteins to be used within the cell, the ribosomes are located in the cytoplasm. For the synthesis of proteins to be exported from the cell, the ribosomes attach themselves to the surface of the endoplasmic reticulum, a specialized tubular structure within the cell. Just as described for DNA replication and transcription, DNA translation also follows three basic steps: initiation, elongation, and termination.

Initiation begins with the binding of a mRNA molecule to a small ribosomal subunit. Each mRNA molecule has a special sequence called the start codon (a sequence of adenine, uracil, and guanine nucleotides). It is a sort of "flag" that indicates where the translation process must begin. Once this sequence has been detected, a special RNA called initiator tRNA attaches. This tRNA carries with it the amino acid methionine, that forms the first amino acid of the new protein. Once the initiator tRNA has joined up with the mRNA and the small ribosomal subunit, a large ribosomal subunit joins, too, and completes the protein building machine. When all these components are in place, initiation is complete and elongation can begin.

During each step of elongation, one amino acid is added to the growing protein. For each amino acid added, the ribosomal complex moves down the mRNA molecule by exactly three nucleotide bases.

- **Each sequence of three nucleotides codes for exactly one amino acid. This code that matches each three nucleotide sequences, or codons, to one amino acid, is called the genetic code.**

Each codon matches up with exactly one tRNA, and each tRNA matches up with just one kind of amino acid. Thus, for twenty amino acids there are at least twenty different kinds of tRNA, one for each amino acid. As it turns out, several different codons can code for exactly the same amino acid, which means that the genetic code is slightly redundant. Still, it works very well and precisely translates the sequence of codons in a gene into a sequence of amino acids in a protein. Each time a codon is matched up with its specific tRNA, the amino acid carried by that tRNA is then added to the growing peptide chain by forming a special strong covalent bond called a peptide bond. After each amino acid addition, the ribosomal unit moves down the mRNA to the next codon, and the next amino acid is added. In this manner, the entire mRNA sequence of codons is read in order and a new protein is synthesized (See Figure 7.4).

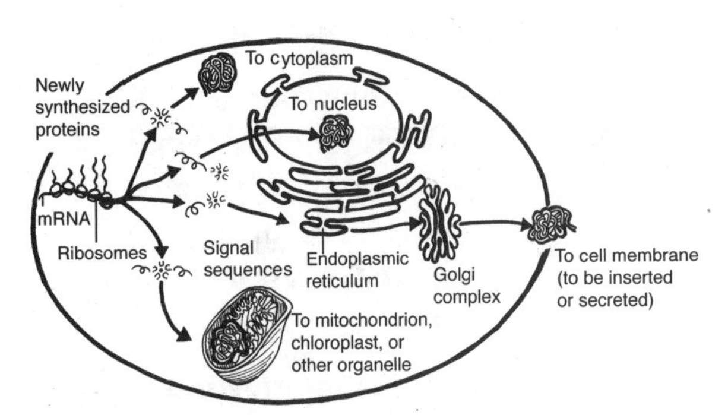

Figure 7.4—Destination of Newly Synthesized Proteins

> Sometimes, when the ribosomal unit has already moved down the mRNA, another ribosomal complex can form on the same mRNA. The addition of these extra ribosomes, called polyribosomes, makes it possible for a single mRNA molecule to simultaneously direct the synthesis of multiple copies of a protein.

At the end of each mRNA molecule are special sequences of nucleotides called stop codons. These codons usually begin with uracil (uracil-adenine-guanine, uracil-guanine-adenine, and uracil-adenine-adenine), and signal the release of the polypeptide, tRNA, and ribosomal subunits from the mRNA. The polypeptide can then be further modified by the cutting of a few amino acids, the addition of sugars or lipids, or the addition of carbon or phosphate containing side groups. Many of these modifications help determine how the protein folds, combines with other proteins, and forms its unique shape. Other modifications regulate how the protein is transported and localized within the cell. Altogether, these post-translational modifications are critically important for regulating how and where a protein functions within the cell.

RECOMBINANT DNA AND BIOTECHNOLOGY

A DNA molecule that contains DNA sequences derived from different biological sources that have been joined together in the laboratory is called recombinant DNA. Recombinant DNA is formed by inserting (splicing) the gene(s) in question into a larger molecule of DNA. Viruses and bacterial plasmids are the usual vehicles for carrying these foreign genes into a host cell. The possibilities of recombinant DNA are almost endless, since the molecule itself, is composed of the same chemicals in all organisms.

There have been many products developed with this type of technology. Vaccines have been made that stimulate the body's immunity to protect against disease-causing viruses and bacteria. Interferons fight viral infection

> DNA is often a silent witness in the criminal justice system. A technique called the polymerase chain reaction (PCR) is being used to enzymatically amplify DNA. This technique can amplify specific fragments of DNA, as well as generate large amounts of DNA from minute starting samples. This procedure has been used in crimes to convict or clear an individual. PCR can generate quantities of DNA from a spot of dried blood left on clothing, etc. Enough DNA for PCR can be obtained from a single hair follicle left at a crime scene.

and boost the immune system. Erythropoietin stimulates red blood cell production. It is being used to combat anemia and increase blood cells in individuals receiving chemotherapy for various cancers. Pigs have been made leaner due to foreign growth hormones. Mice have been made larger by carrying the rat growth hormone.

SUMMARY

- The discovery of DNA as the hereditary material took the efforts of many scientists. Hershey and Chase were able to convince the scientific community that it was DNA, and not proteins, that was the genetic material. Watson and Crick were able to finalize the structure of DNA by interpreting all of the scientific data on this molecule.
- DNA is a double helix composed of a backbone of deoxyribose, a phosphate group, and nitrogenous bases. These bases are bonded by hydrogen bonds; adenine always pairs with thymine and guanine always pairs with cytosine.
- DNA replication is semiconservative. The strands of the helix unwind, and each strand serves as a template for a new strand of DNA. Each gene consists of a linear sequence of nucleotides that determines the linear sequence of amino acids in a protein. The processes of transcription and translation accomplish this.
- During transcription, the information in the gene's sequence is encoded in a molecule of mRNA. The mRNA is complementary to DNA and leaves the nucleus after modification of the molecule. The mRNA consists of codons—each is a sequence of three bases. All of the codons specify amino acids, except UAG, UGA, and UAA, which code for termination of the growing polypeptide chain.
- During translation, the codons of mRNA are used as the basis for the assembly of specific polypeptides. Translation occurs on the ribosomes and requires tRNAs, which bring over the necessary amino acids.
- Messenger RNA is an information carrier. It carries the genetic information from DNA to the ribosomes. The message specifies a protein's primary structure. Transfer RNA acts as an adapter in protein synthesis by translating information from one (mRNA nucleotide sequence) into another (protein amino acid sequence). Ribosomal RNA plays a structural and probably enzymatic role in the ribosomes.
- The biotechnology revolution has been built on the development of new techniques in DNA technology. Recombinant DNA has given us many new products that are useful. The role of PCR in forensics is firmly established.

CHAPTER 8

EVOLUTION

KEY TERMS

zygote	adaptation	
niche	biodiversity	mutation

Why don't we all look alike? Why do some animals have horns and others have claws? Why do some species die out and others survive? Charles Darwin probably pondered these same questions. His answers are a basis for modern biology. When his ideas first came to public attention, they had a tremendous impact on the world of science. Darwin's theory has been refined, developed, tested and studied beyond his imagining.

PRE-DARWIN IDEAS

From about 300 BC, Greeks theorized that an organism's form was related to its function. This was the prevailing belief until 1769, when Swiss naturalist, Charles Bonnet, proposed that periodic catastrophes affected the entire planet, and after each catastrophe, life began again. He was the first to use the word evolution to explain the development of diverse life forms. In 1809, Jean-Baptist Lamarck theorized that single organisms emerge spontaneously and evolve to greater complexity. He was later disproved since change in an individual during its lifetime doesn't affect reproductive cells or offspring.

CHARLES DARWIN

H.M.S. *Beagle*

In 1831, Charles Darwin began a five-year journey on the H.M.S. *Beagle* as a naturalist (See Figure 8.1). Darwin took copious notes and collected specimens around the world, returning to London where he continued his studies. In 1844, he wrote his theory on the origin of the species. It states, simply, that evolution does occur; evolutionary change is gradual, requiring thousands to millions of years; evolution occurs through natural selection; and millions of species arose from a single life form through specialization.

Before Darwin set foot on the H.M.S. *Beagle*, most people believed that plant and animal species were unchanging. It was more than ten years before he published his ideas, knowing that they would certainly be challenged. His book, *On the Origin of the Species by Means of Natural Selection*, was published in 1859.

Darwin's ideas were met with suspicion by many in nineteenth century England. The concepts to challenged the religious beliefs of the time that advocated the theory of creation, and mankind's supremacy in the natural world. Although this conflict continues to be an issue today, evolutionary theory has been rigorously tested. While it may be impossible to prove it to be correct, it has never been disproved.

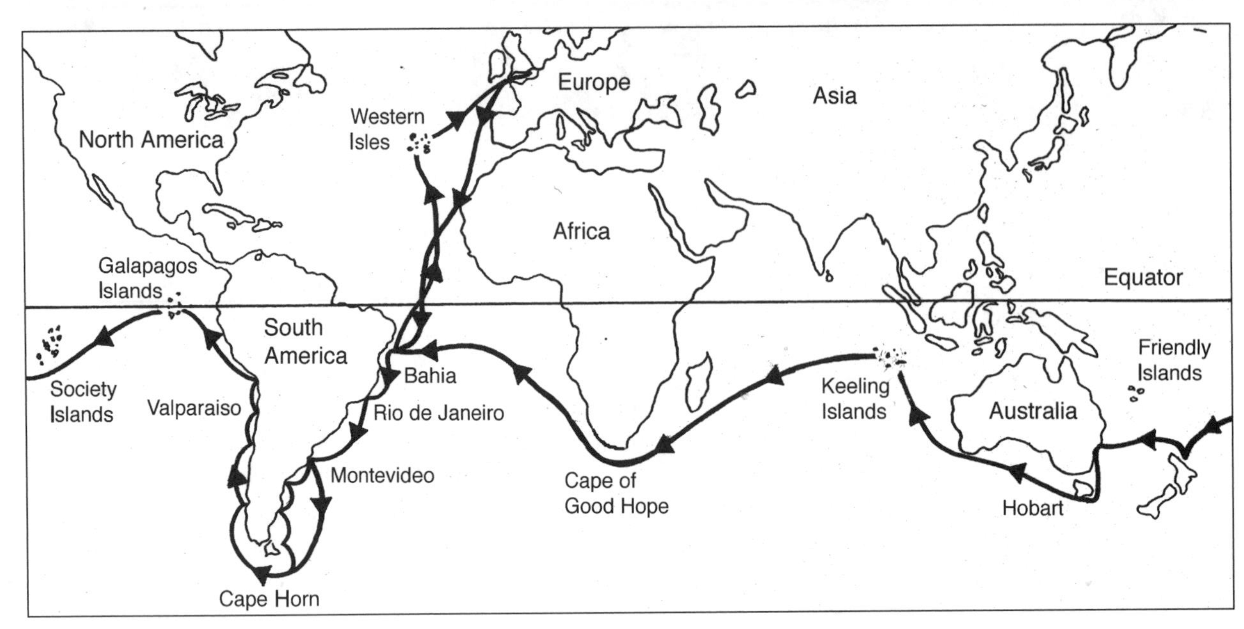

Figure 8.1—Darwin's Voyage on the H.M.S. *Beagle*

- **English biologist, Alfred Wallace, developed a theory similar to Darwin's. The two agreed to send their papers out at the same time in 1858, with Darwin having credit for developing the theory first.**

Natural Selection

The key to evolutionary theory is natural selection. As Darwin traveled around the world, he was fascinated, among other things, by the finches that he observed on the Galapagos Islands off the western shore of South America. The different types of finches were later found to each belong to a separate species. They did not interbreed. Each type of finch, though living in close proximity, had adapted characteristics that suited its own environment.

A species is a reproductively isolated population (made up of individuals that breed solely within the specific group). The organism in a population that is well adapted to a particular environment is going to have an advantage over other organisms. It is more likely to survive, be reproductively successful, and pass on its genetic material. For example, let's say that on one island, birds with long beaks are better able to obtain food. Beak length would be an advantage in its environment, and would become an important criterion by which mates are selected. An individual's ability to reproduce and pass on desirable traits was as important as survival itself.

Natural selection is based on four important ideas:

- There is variation within populations.
- Some variations are favorable.
- Not all young produced in each generation can survive.
- Individuals that survive and reproduce are those with favorable variations.

The observation that the finch populations lived close together on the tightly grouped Galapagos Islands, and could have interbred, but didn't, made Darwin curious. What isolated one group from another reproductively?

Prezygotic and Postzygotic Isolating Mechanisms

Prezygotic isolating mechanisms are factors that take place before fertilization occurs. Geographical isolation becomes a factor when geographical barriers keep groups isolated. In ecological isolation, two subpopulations living in different habitats in the same area may fail to meet at breeding time. A shift in the time of flowering for a plant is an example of seasonal isolation. Mechanical isolation can include structural differences in the sex organs. Finally, behavioral isolation may be the result of differences in mating behavior.

Even if fertilization occurs, postzygotic isolating mechanisms may prevent reproduction. The embryo might form, but the genetic differences may be so great that they fail to reach reproductive maturity.

- **Hybrid sterility means fertilization is successful, but the offspring is sterile. An example is the mule produced from the mating of a horse and donkey.**

THE BASIS OF EVOLUTIONARY CHANGE

Not all change is evolution. For example, a person may alter her appearance through diet and exercise. Someone may change his looks through surgery. Influences such as these may lead to change, but the changes are not passed to future generations and are not an example of evolution. Phenotypic changes in a population over time are evolution in action. They are due to genetic change and they can be passed to the offspring.

Mutations

When the cellular machinery that copies DNA makes a mistake and alters the sequence of a gene, it is a mutation. One "letter" of the genetic code may be changed to another, lengths of DNA can be deleted or inserted into a gene, and genes or parts of genes can be inverted. A mutation creates new alleles, or forms of a trait, which enter the gene pool. Sometimes the trait that results is beneficial and gives the organism some type of advantage. Over time, such a trait will show up more often in the population.

Most mutations with phenotypic (observable) effects are not beneficial, such as an extra digit on the hand. Most of these mutations are lost, either because the individual carrying it is considered undesirable or because the trait is recessive and is not passed on.

Recombination

The genetic makeup of a species can change through recombination. During meiosis (see Chapter 4) chromosomes are taken apart and recombined. In humans, twenty-three restructured chromosomes make up the sex cells or gametes. When egg and sperm join, the resulting zygote has a full set of forty-six chromosomes, half from the male and half from the female. The individual developed from that cell is unique. Recombination can occur between or within genes. Within genes it can add a new allele to the gene pool, opening a path to evolution.

- **The gene pool is the combined genetic material of all the members of a given population. It includes all the genes for every trait on every chromosome of all the reproductive individuals in that population.**

Flow and Drift

Gene flow occurs when new organisms migrate into an existing population. Mating within the blended population adds new alleles to the local gene pool. Genetic drift occurs when allele frequencies change due to chance alone. Evolution often begins with changes in allele frequencies over time.

Both natural selection and genetic drift decrease genetic variation. If these were the only factors at work in evolution, all populations would become the same, and evolution would cease. There are many other ways that life forms can replace variation that is depleted by either natural selection or genetic drift.

- **Although the overall rate of genetic drift is independent of population size, since there is less to work with a smaller population, the impact is greater.**

Genetic Selection

Stabilizing Selection

Natural selection affects a population in several ways. The first of these is stabilizing selection. This process eliminates individuals at both extremes, meaning that those with the most reproductive success will be somewhere near the middle, thus maintaining the status quo.

This may not exactly sound fair, but it is common. An unfortunate example might be the tendency toward infant mortality in babies that are either very heavy or very low weight at birth.

Directional Selection

A population can also be in a situation where individuals occupying one extreme are favored over others. For instance, let's say that there are individual birds within a population of seed-eaters that have beaks a little larger and stronger than those of the general population. Imagine that a severe long-term drought leaves only plants that produce seeds in large, tough fruits. The result is that the birds with weak beaks, and many of the average-beaked birds fail to find enough available food. They may die off and so fail to reproduce. The birds with larger beaks reproduce and that trait is passed to future generations. Once the drought ends, the population increases again, but large-beaked birds have gained the edge.

Disruptive Selection

When individuals at both extremes of a range of phenotypes are favored over those in the middle, it is called disruptive selection.

For instance, some hardy species, such as certain grasses, might be able to grow on the site of mines with contaminated soil. They can develop resistance to the toxic metals while their ability to grow in uncontaminated soil decreases. Because grasses are wind pollinated, breeding between the resistant and nonresistant populations goes on. However, there are higher death rates of both the less resistant plants on contaminated soils and the more resistant plants growing on uncontaminated soil, which leads to an increasing divergence of the populations into two subpopulations with extreme manifestations of their respective trait. There is now the possibility that the gene pool may be

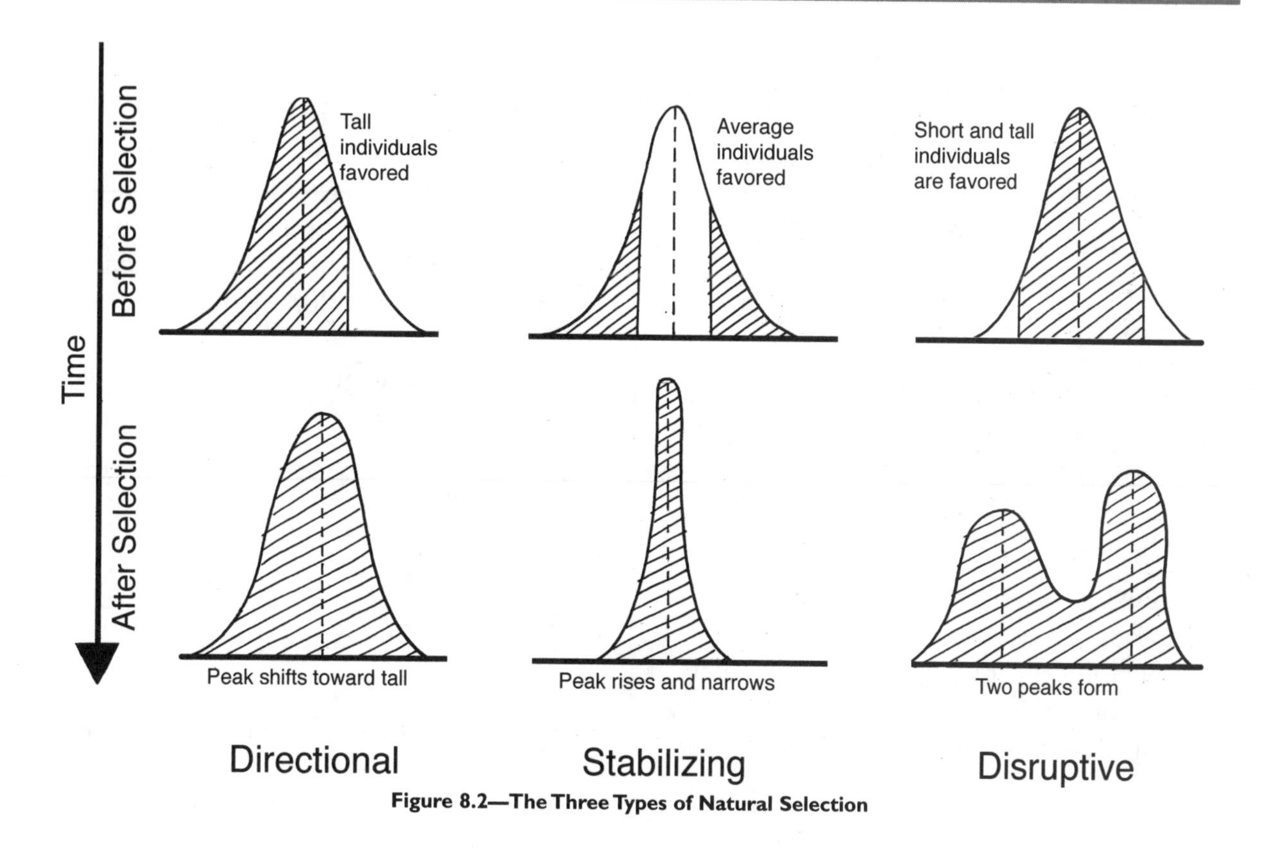

Figure 8.2—The Three Types of Natural Selection

split into two distinct gene pools and a new species could eventually form.

HARDY-WEINBERG PRINCIPLE

In 1908, three mathematicians, (Castle, Hardy and Weinberg) proved that random mating alone could not change allelic frequencies. The Hardy-Weinberg Principle states that allele frequencies in a population remain constant from one generation to the next. When a large population must mate at random, there is no migration, no mutation, and no natural selection. This theory was important because it showed that random mating alone does not cause evolution.

On the other hand, evolution does result from *nonrandom mating,* also called *assortive mating.* This limits breeding to members of a population that possess certain desirable traits, such as the strongest male bear or the male bird with the yellow crest.

When females choose males that "win" competitions (in deer, for example), a few males pass on genes to the next generation. According to Darwin, this was sexual selection and it could cause evolution.

PATHS OF SPECIATION

The formation of one or more species from a single species, or the increase of biological diversity, is called speciation. Some biologists believe that this is the key to understanding evolution. There are three basic types:

- **Allopatric speciation** is the evolution of reproductive isolating barriers that occur as the result of an outside, or physical, isolating barrier. It happens when a population is split into two or more isolated groups and the organisms cannot physically reach each other. The two populations' gene pools change independently until they could not interbreed even if they were brought back together. In vicariance, a geological or physical barrier causes the isolation, and in a founder event, a rare colonization of a new area occurs with a limited connection to the parental population.
- **Parapatric speciation** is the evolution of reproductive isolating barriers between geographically localized subsections of a continuously distributed species. It usually happens as a result of an abrupt change in the environment.
- **Sympatric speciation** is the evolution of reproductive isolating barriers within a randomly mating population. In other words, the two subpopulations become reproductively isolated without first becoming geographically isolated. An example of this might be insects living on a single host plant. If a group switched plants, though still in close proximity, they would not breed with other insects still living on the original host plant.

Hybridization can occur when two genetically different populations or species come into contact, reproductive isolating barriers are incompletely formed or weak, and mating occurs. The two populations are genetically crossed to produce some of the characteristics of the "parents," plus some new characteristics.

PATTERNS OF EVOLUTION

Natural selection, through the various processes mentioned, can lead to the formation of new species. This can occur in a variety of ways, including divergent evolution, adaptive radiation, convergent evolution, parallel evolution and coevolution.

- **Adaptive radiation** is the evolutionary pattern that occurs when many species evolve from a single ancestral species. It usually occurs when a species of organisms successfully invades an isolated region where few competing species exist. If new habitats are available, new species will evolve suited to each niche.
- **Divergent evolution** is when two or more related species become more and more dissimilar. An example of this might be the red fox and the kit fox. The red fox blends into farmlands and forests with its red color. The kit fox, with its sandy color, lives on plains and in the deserts. Similarities in structure indicate that the two foxes had a common ancestor, but as they adapted to different environments, the two species diverged.
- **Convergent evolution** takes place when two unrelated species become more and more similar as they adapt to the same kind of environment. Unrelated types of plants that exist in the desert are an example of this. Fleshy stems armed with spines help plants store water and ward off predators. These adaptations show up on a variety of desert plants even though they are not the same species (See Figure 8.3).

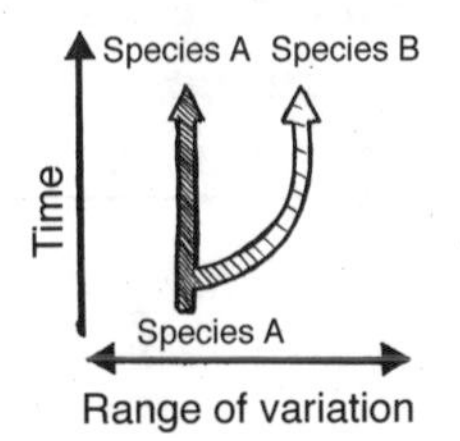

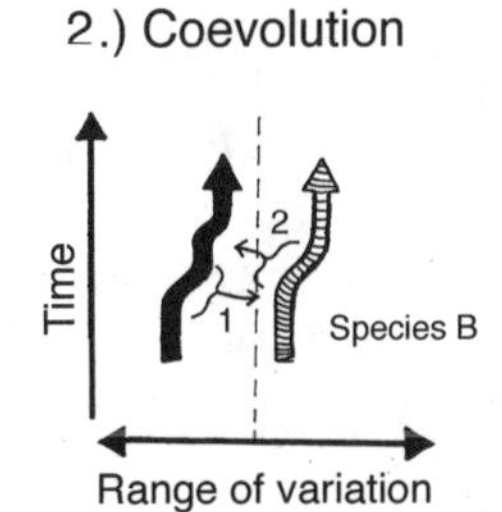

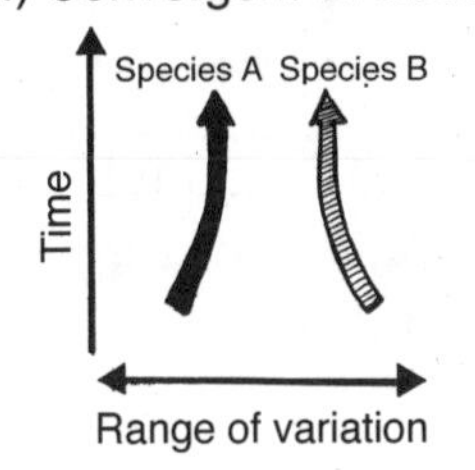

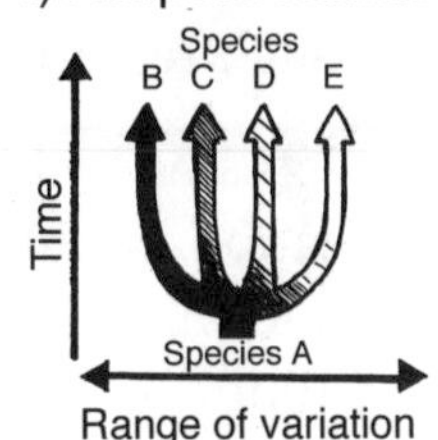

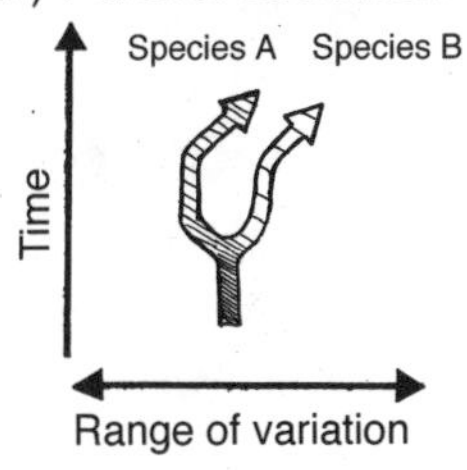

Figure 8.3—Patterns of Evolution

- **Coevolution** occurs when two or more species in close interaction undergo joint change, for instance between plants and the animals that pollinate them. Bats, in tropical regions, eat nectar from flowers. The fur on the bat's face and neck picks up pollen, which the bat transfers to the next flower it visits. The bats have slender muzzles and long tongues with brushed tips. These adaptations aid the bat in feeding. Flowers that have coevolved with bats are light in color so bats can easily locate them. The flowers also have a fruity odor attractive to bats.

- **Parallel evolution** is the development of the same characteristics or adaptation in unrelated organisms due to similar environmental conditions. They don't necessarily occupy the same niche in a habitat, but evolve independently, maintaining a level of similarity. An example is the python of Asia and the boa constrictor of South America.

EXTINCTION

Scientists estimate that there are 5 to 30 million different species on earth today, and yet ninety-nine percent of all species that ever lived are now extinct. Extinction is decreased biological diversity brought about by a variety of factors. Perhaps the habitat of a species disappears. A predator may develop or migrate into an area where the prey are vulnerable. A shift in sea level might allow mixing of species. One species might lack defenses to a disease carried by another.

> Some species enjoy a long tenure on the planet while others are short-lived. If the environment stays fairly constant, a well-adapted species could continue to exist indefinitely—certain insects, for example.

The history of life on earth is punctuated with episodes of mass extinction, such as the end of the Permian, 250 million years ago. This coincides with the formation of Pangaea II, when all the world's continents were brought together by plate tectonics. The shift resulted in a worldwide drop in sea level.

About 65 million years ago, at the boundary between the Cretaceous and Tertiary Periods, dinosaurs were eradicated. This may have been caused by some massive environmental disruption and opened a window of opportunity for small mammals that had coexisted with dinosaurs. Eventually they filled the niches left by the dinosaurs—and the age of mammals began.

SUMMARY

- In the mid-nineteenth century, Charles Darwin traveled on the H.M.S. *Beagle,* then published his famous book, *On the Origin of the Species by Means of Natural Selection.* His ideas revolutionized scientific and popular thought when he introduced the idea that species evolved over time.
- Natural selection showed that there was variation in populations, some of which was favorable. It was the individuals with favorable characteristics who survived and passed on those traits through reproductive success.
- Darwin's studies also established that mechanisms occurring both before and after fertilization can predetermine whether or not an organism is viable.
- Evolutionary change takes place through mutations, recombination, gene flow, and genetic drift.
- Speciation is the formation of one or more species from a single species through reproductive isolating barriers.
- Patterns of evolution include adaptive radiation, divergent evolution, convergent evolution, coevolution, and parallel evolution.
- Numerous species have become extinct over millions of years. At the same time, new species evolve. Biodiversity is the abundance and variety of life forms and while some species become extinct, others evolve. Evolution guarantees the variety and abundance of species that make up our biological community on earth.

THE DIVERSITY OF LIFE

KEY TERMS

pathogen	capsid	capsomere
vector	plasmodium	hyphae
xylem	phloem	

Humans have walked on the moon, conquered numerous diseases, and learned to control the elements in many ways. Animals have evolved that can race at remarkable speeds, fly in the air, and swim in the sea. The road to these achievements began about three billion years ago with tiny, single-celled organisms called prokaryotes.

Prokaryotes were essential to the earth for several reasons. Some employed a primitive form of photosynthesis. Others were decomposers that could break down organic materials, which recycled nutrients back into the environment.

ARCHAEBACTERIA AND EUBACTERIA

Prokaryotes are separated into archaebacteria and eubacteria because of dramatic differences in gene sequences between the two groups. In turn, archaebacteria are grouped into three phyla of bacteria found mainly in extreme habitats where little else can survive. They might be found at the bottom of the sea or in volcanic vents. They are generally anaerobic (they live without oxygen).

- **Methanogens** are the type of bacteria that produce methane. They live swamps, sewage, and land fills.
- **Halophiles** can only live in bodies of concentrated salt water, such as the Great Salt Lake in Utah, the Dead Sea in the Middle East, or other areas with a high salt content.
- **Thermoacidophiles** live in the hot, acidic waters of sulfur springs. They can survive in areas with temperatures as high as 230 degrees Fahrenheit and with a pH below 2.

Eubacteria live in a wider variety of habitats than archaebacteria. They reside in our bodies, on our food, and in our homes.

- **Heterotrophs** might live as parasites, absorbing nutrients from living organisms, or as saprobes, organisms that feed on dead organisms or organic waste. Their role in decomposition is important in recycling nutrients, which can be used for new or existing life.
- **Autotrophs,** another form of eubacteria, can make food through photosynthesis. They live in ponds, lakes, streams, and other moist areas. Because they are composed of chains of bacteria cells, an exception to the rule that prokaryotes are unicellular, they help provide proof that bacteria are the ancestors of plants.
- **Proteobacteria** is one of the largest phyla of all bacteria. They are divided into several

CLASSIFICATION CRITERIA FOR THE THREE DOMAINS

	Domain Bacteria and Archaea	Domain Eukarya			
		Kingdom Protista	Kindgom Fungi	Kingdom Plantae	Kingdom Animalia
Type of cell	Prokaryotic	Eukaryotic	Eukaryotic	Eukaryotic	Eukaryotic
Complexity	Unicellular	Unicellular usual	Multicellular usual	Multicellular	Multicellular
Type of nutrition Heterotrophs	Autotrophs or heterotrophs	Photosynthetic or heterotrophs		Heterotrophs	Photosynthetic (by ingestion)
Motility	Sometimes by flagella	Sometimes by flagella	Nonmotile	Nonmotile	Motile by contractile fibers
Life cycle	Asexual usual	Various	Haplontic (asexual)	Alternation of generations	Diplontic (sexual)
Internal protection of zygote	No	No	No	Yes	Yes
Nervous system stimuli	None	Conduction of	None	None	Present

subgroups such as enteric bacteria, chemoautotrophs, and nitrogen-fixing bacteria.

- **Chemoautotrophs** are chemosynthetic autotrophs that obtain their energy from chemosynthetic breakdown of substances such as sulfur and nitrogen compounds. These are essential to many ecosystems and one type converts the unusable nitrogen in the atmosphere to ammonia, the form plants can use most easily.
- **Enteric bacteria,** *E. coli* for example, live mainly in the intestinal tracts of animals.

Forms of nutrition and nearly all metabolic pathways evolved among prokaryotes before eukaryotes arose. This change probably happened in steps, the first of which was for one reaction to utilize an end product of another reaction. Over the past 3.5 million years, eukaryotes have developed many elaborate ways of doing this same basic process. Some processes, like photosynthesis, may have evolved to protect the organism from toxic substances in the first place.

- **Living things are classified into kingdoms. Many scientists use a higher classification in which kingdoms are ordered in three domains: Archaea, Bacteria, and Eukarya.**

Morphological Diversity

Beyond the classifications that group bacteria by habitat and means of existence, they are also grouped by gram staining. Gram staining involves staining a bacterium with four different liquids, crystal violet, then an alcohol wash, iodine, and then safranin. The bacteria that retain the purple stain from the crystal violet are gram-positive, and those that take on the pink from safranin are gram-negative.

MAJOR DIFFERENCES AMONG THE THREE DOMAINS

	Bacteria	Archaea	Eukarya
Unicellularity	Yes	Yes	Some, many multicellular
Membrane lipids	Phospholipids, unbranched	Varied, branched lipids	Phospholipids, unbranched
Cell wall	Yes	Yes	Some yes, some no
Nuclear envelope	No	No	Yes
Membrane-bounded organelles	No	No	Yes
Ribosomes	Yes	Yes	Yes
Introns	No	Some	Yes

Gram-positive bacteria have a thick layer of peptidoglycan (an amino acid compound linked to sugar), which absorbs the gram stain. Gram-negative bacteria have a thick lipid bi-layer on the outside, which is selectively permeable.

- **Since substances pass easily through the outer layer of their cells, gram-positive bacteria are more susceptible to antibiotics than gram-negative bacteria.**

Bacteria are also classified by their shape—bacilli, cocci, and spirilla. Bacilli are oblong and thick. Cocci are round, and spirilla are spiral-shaped. They also have different ways of moving: by a corkscrew motion, by gliding, or by using flagella that propel them along.

PATHOGENS

Some bacteria cause disease. Scientists have found that a type of bacteria called cocci occuring in chains (streptococi), or in bunches like grapes (staphylococci) are common causes of infection. An organism that causes an infectious disease is called a pathogen. Viruses, bacteria, protists, fungi and some invertebrate animals such as roundworms and flatworms can be pathogenic to humans.

Some pathogens produce toxins that disrupt the normal functioning of cells (for example, the pathogens that cause diphtheria, cholera, and pertussis). Viruses, such as chicken pox and the common cold, use the body's cells to replicate. Protists and roundworms destroy body tissues by feeding on and burrowing into the tissue. Humans, however, are equipped with an immune system that seeks out and often destroys pathogens.

VIRUSES

Viruses are small obligate intracellular parasites, which may contain either an RNA or DNA genome surrounded by a protective, virus-coded protein coat. Viruses depend on specialized host cells that supply the complex metabolic and biosynthetic machinery the viruses need to reproduce. Viruses are not cells and viral particles do not have a nucleus, a membrane, or cellular organelles such as ribosomes, mitcohondria, or chloroplasts. They do, however, have organized structural parts—a protective protein coat (capsid) and a core of nucleic acid. A human cell may contain about 100,000 genes, while a bacterial cell has about 1,000 and a virus may contain only five genes.

- **Viruses, unlike cells, do not eat, respire, or respond to environmental changes.**

Viruses are grouped into 21 families on the basis of size, shape, chemical composition, structure of the genome and mode of replication. The number and arrangement of the capsomeres making up the protective coat, or capsid, are useful in identification and classification. The genome may occupy either one nucleic acid molecule or several nucleic acid segments, and the different types of genomes necessitate different replication strategies.

Emerging and Reemerging Infections

Antiobiotics have been seen as miracle drugs in the past century, curing many types of common infections that once killed large numbers of people. This does not justify complacency, however, as emerging and reemerging infections prove resistant to antibiotics that once seemed guaranteed to work. The human immunodeficiency virus (HIV) has proven to be one of science's greatest modern challenges. There have also been cases of multi-drug resistant tuberculosis, acute coccal infections, rodent-borne penumonic hantavirus, and food and waterborne outbreaks of *Salmonella* and other infections caused by *E. coli.*

Because the human population is more mobile than ever, infectious diseases know no boundaries. People travel everywhere, carrying viruses in a variety of ways. Many viruses travel via rodents and arthropods, so changes in agricultural practices or urban conditions can have a significant impact for the worse. Deforestation, irrigation, long-distance livestock transport, and rerouting of long-distance bird migration brought about by man-made water control, can all affect virus travel and growth.

One of the most significant examples of how ecological changes have had an impact on the growth of a virus might be in dengue. This disease is common in tropical areas, and there are millions of cases each year. It has spread into a number of new countries in recent years. *Aedes albopictus*, an aggressive dengue virus vector, was brought to Houston in used Asian tires and has established itself in a number of American states.

Viroids and Prions

There are two infectious agents with simpler structures than viruses: viroids and prions. A viroid is a single strand of pure RNA, which causes plant diseases. Unlike viruses, viroids do not have capsids protecting their nucleic acids.

Prions are protein molecules that can cause disease in animals. They are the only known infectious agents that do not contain DNA or RNA but can still spread throughout an organism. Prions have been found in the brains of cows that died from mad cow disease, and in humans who suffer from kuru or Creutzfeldt-Jakob disease. Both diseases affect the central nervous system.

PROTISTS

Protista are unicellular and multicellular eukaryotes. These cells have nuclei and organelles that are surrounded by membranes. Most are not specialized.

The animal-like protists (protozoans) differ from animals in that they are unicellular and do not have specialized tissues, organs, or organ systems that carry out life functions. They are classified into four groups according to how they move.

- **Sarcodinians** move by extending lobes of cytoplasm (cell interior).
- **Zooflagellates** use flagella, or tendrils, in order to move. Some use the flagella to move in a relatively straight path, others create currents that spin them through fluids.
- **Cliliophorans** are the most numerous and diverse group of protozoans. They are covered with cilia, short, hairlike projections used for movement. Most ciliates live in freshwater habitats and can form cysts to survive in unfavorable environments.
- **Sporozoans** are parasitic protozoans. They have no structures used for movement. The life cycles of sporozoans are complex and include both sexual and asexual phases. The life cycle often involves more than one host, such as when plasmodium infects both mosquitoes and humans, causing human malaria.

Algae

The protists that perform photosynthesis are called algae. Algae are like most plants in that they contain chlorophyll and produce food and oxygen as a result of photosynthesis, but algae does not contain specialized tissues or organs. The best way to organize algae for study is based on their differences in structure: unicellular and multicellular.

Single-celled algae include dinoflagellates, diatoms and euglenoids. Dinoflagellates have two flagella which spin their cell in a corkscrew fashion through water. Cellulose plates cover each dinoflagellate. Diatoms look like snowflakes and have glasslike cell walls containing silica. Pores in the cell walls allow material to pass into and out of the diatom. Euglenoids resemble both algae and protozoans. They lack rigid cell walls and they move using flagella, but they have been classified as algae because they have chloroplasts and perform photosynthesis.

Multicellular algae have specialized structures like plants, but are classified as protists because very few have true tissues, and their reproductive methods are more like those of protists than plants. Multicellular algae are classified by color, determined by the pigment in their cells. There are green, red, and brown algae, all of which live in different habitats.

Slime Molds and Water Molds

Fungus-like protists are molds. They tend to be small, live in damp or watery places, and act as decomposers. Molds are divided into three groups—plasmodial slime mold, cellular slime mold, and water mold.

The plasmodial slime mold is a single cell with many nuclei. It can grow as large as a hamburger. When circumstances are not conducive for this slime mold it forms a structure called a fruiting body, which produces spores that are distributed by wind and animals. Under favorable conditions, the plasmodium releases haploid gametes that fuse to form a diploid zygote, which develops into a new plasmodium.

- **Slime molds get their name from their shiny, wet appearance, and a texture like gelatin.**

Cellular slime molds alternate between a spore-producing, fruiting body and an ameoba-like feeding form. When conditions are unfavorable, the cells secrete a chemical attractant that causes nearby mold cells to clump together to form a pseudoplasmodium, a clump of cells that produces fruiting bodies. The cells, however, are independent entities and are haploid.

Water molds are decomposers in freshwater ecosystems. Some are parasitic and attack the injured skin or gill tissues of fishes and others are parasites of certain land plants. Watermolds can't be mistaken for fungi because they have cell walls that are mostly cellulose, as opposed to chitin (a polysaccharide). Also, asexual reproduction in water molds produces spores with flagella and fungi produces spores without flagella.

FUNGI

Fungi were once classified as members of the plant kingdom, but they differ from plants in several important ways. The cell walls of plants are mostly cellulose, while that of fungi is chitin. Fungi cannot make their own food.

Fungi are classified in their own kingdom but the main groups are called divisions rather than phyla. These groups are based primarily on their method of reproduction. Common molds, sac fungi, and club fungi are classified based on the structures used in sexual reproduction, and the fourth group, imperfect fungi, are those that reproduce only asexually.

Actually, all fungi can reproduce asexually, through budding, regeneration or spore production. Most fungi reproduce asexually by producing spores in a structure called a fruiting body.

Common molds, sac fungi and club fungi can also reproduce sexually. Rather than combining the genes of a male and female parent, fungi have two different mating types of hyphae (branching, tubular cells) plus (+) and minus (–). Sexual reproduction occurs by the fusion of two hyphae with different mating types, and the production of spores. The spores that are produced sexually contain a new combination of genetic information.

Some fungi depend on living hosts for food. They can be parasites, mutualistic partners, or predators. The parasitic fungi will absorb nutrients, often causing disease or the death of the host. Fungi that live in a mutualistic relationship absorb nutrients from the host, but they also provide the host with necessary materials, such as minerals from the soil. Some fungi are predators that trap and kill their prey.

Economic Importance

Fungi are versatile and can save your life, kill you, and everything in between. Mushrooms are an example of the dangers and benefits of fungi. Some common mushrooms are poisonous, and others are extremely delectable. An imperfect fungus, *aspergillus* is used to make citric acid and soy sauce, while yeast has been used for centuries to make bread rise, and to ferment beverages.

One of the most critical uses of fungi involves *Penicillum*, a fungal mold that rots fruit. When scientists discovered that this mold killed the bacteria *Staphyloccus*, it proved to be a life-saver for humans. Today, antibiotics are synthesized.

- **Genetically altered yeasts are used to synthesize many important proteins and are critical in genetic engineering.**

PLANTS

Plant Structure and Growth

The first stages of plant growth take place at the extremities. Growing tissues, called meristems, are located at the tips of stems and branches, at the tips of roots, and in the buds in the joints where leaves attach to stems. Their function is to produce new cells by mitosis. These cells, alike at first, eventually specialize to become vascular tissue, dermal tissue, or ground tissue. Apical meristems cause the roots and stems to grow longer, and axillary meristems are in the buds that arise in the joints between leaves and stems.

The elongation of stems and roots is primary growth, during which the plant grows taller, and its roots sink deeper. Secondary growth is when it grows wider. Although all plants experience primary growth, not all have secondary growth. Lateral meristems shaped like hollow cylinders inside a root or stem are necessary for secondary growth. These are also called cambium and can be either *vascular* or *cork*. Vascular cambium is between the xylem and phloem (fluid transport structures), and cell divisions in the vascular cambium produce new xylem toward the inside of the cambium, and new phloem toward the outside of the cambium. This growth takes place in a cycle.

Cork cambium is between the phloem and the epidermis. Cell divisions in the cork cambium replace the cortex layer and the epidermis, or "skin," of the plant.

Xylem accumulates yearly, but only the newer outer layers continue to transport water. Older xylem cells get clogged. You can see this growth process by looking at the growth rings in a cross-section of a tree.

Reproduction and Biotechnology

Because plants can't move around, sexual reproduction is a complicated process which requires that sex cells meet. All plants undergo alternation of generations. They switch from diploid generations to haploid generations. The diploid plant has two complete sets of chromosomes. Instead of producing gametes, the diploid plant undergoes meiosis to produce haploid spores. Plant spores divide by mitosis to form a plant structure in which each cell is haploid.

The haploid plant produces gametes in specialized structures by further mitotic divisions. During fertilization, some of the haploid gametes will fuse with other haploid gametes, forming a zygote, which develops into a new diploid organism. The new diploid plant has two sets of chromosomes, one from each haploid parent.

Responses to Internal and External Signals

Plants respond to conditions in their internal and external environments. A plant may respond to light, gravity, or touch. Plants produce small quantities of chemicals that control growth and response. Seasonal changes determine the timing of bud and flower production. Plants usually become inactive during periods of extreme cold or in dry conditions, and many seeds and bulbs are inactive for a time before they sprout.

ANIMALS

Invertebrates

Animals without any backbone are invertebrates. They are cold-blooded, which means that their body temperature is regulated externally. Some common invertebrates include sponges, arachnids, insects, crustaceans, mollusks and echinoderms.

Mollusks

Mollusks include snails, slugs, scallops and squid, and although they all look different, they share important structural characteristics. Mollusks have a soft body that may be protected by a hard shell. Some have a single shell. Others have a two-piece shell. Squids and slugs have a light internal shell, and octopuses have no shell at all. Nonetheless, all mollusks have a mantle, a soft outer tissue layer. The mantle often produces a protective shell, and a mantle cavity forms between the mantle and underlying tissue. The mantle cavity houses a mollusk's respiratory structures.

The second common characteristic is a muscular foot that is used to move or hold onto something. Between the foot and mantle is the visceral mass (which contains the main organs of the body), the third important characteristic of a mollusk. The final characteristic shared by many mollusks is an open circulatory system. The blood actually flows out of vessels and bathes the body organs before draining into small vessels and continuing through the respiratory structures. Gas exchange occurs in the respiratory structures as water or air moves over them and from here the blood moves back into the heart.

Annelids

Annelids are earthworms, leeches and other worm-like animals whose bodies are divided into segments. They have body cavities completely lined with mesoderm (a tissue layer). Muscles and fluid in the coelom (body cavity)

SOME CHARACTERISTICS OF EUKARYOTES

Kingdom	Cell Type	Cell Number	Major Mode (of nutrition)	Motility (movement)	Cell Wall	Reproduction
Protista	Eukaryotic	Unicellular	Absorb, ingest or photosynthesis	Both motile and nonmotile	Present in algae forms; Varies	Both sexual and asexual
Animalia	Eukaryotic	Multicellular	Ingest	All motile at some stage	Absent	Both sexual and asexual
Fungi	Eukaryotic	Most multicellular	Absorb	Generally nonmotile	Present: chitin	Both sexual and asexual
Plantae	Eukaryotic	Multicellular	Photosysnthesis	Generally nonmotile	Present: cellulose	Both sexual and asexual

act as a hydrostatic (fluid pressurized) skeleton. The muscles run both lengthwise and around the annelid's body.

- **Annelids also use external structures for movement. Earthworms, for instance, have small bristles called setae.**

Annelids have a well-developed nervous system, and usually a small brain that extends into a solid nerve cord. Earthworms also have receptors in their skin, which are sensitive to light, temperature, moisture, vibrations, and chemicals. They can reproduce asexually by regeneration, but they usually reproduce sexually.

Arthropods

Arthropods are the largest phylum. They are often near the base of the food chain in their respective environments, and sometimes serve as food for other arthropods.

Arthropods have segmented bodies with jointed appendages and a chitinous exoskeleton, which must be molted and shed for growth to continue. Crustaceans, spiders and insects are included in this phylum.

While some insects can cause devastation by damaging crops, hosting dangerous parasites, and carrying disease, others are important to the human food supply. Honeybees pollinate a huge variety of flowering plants and crops, and are valued for their honey and beeswax. Silkworms and crabs have important uses, and people also consume great quantities of shrimp, lobster, crabs, and crayfish.

Echinoderms

Echinoderms are marine animals characterized by spiny skin, an endoskeleton, radial symmetry and a water vascular system. Sea stars, one of the most common echinoderms, are able to reproduce asexually by regeneration, but usually reproduce sexually. They gather at certain times of the year, and then release sperm and eggs into the water, where fertilization occurs. The larvae swim freely for about two years, then settle to the ocean floor and mature into adults.

Chordata

The phylum Chordata contains both vertebrates and invertebrates. Vertebrates—fishes, amphibians, reptiles, birds and mammals—are the most familiar members of the phylum Chordata. The invertebrate chordates, which do not have a backbone, are less familiar. These are the tunicates and lancelets.

Vertebrates

Fish, amphibians, reptiles, birds, and mammals are vertebrates and belong to the phylum Chordata. They share four main characteristics, although these change as the vertebrate develops. The dorsal nerve cord becomes a spinal cord and brain, the notochord is replaced by a backbone. In aquatic vertebrates, the gill slits or pouches become gills. In terrestrial vertebrates, the gill slits or pouches develop into other structures. The post-anal tail is the only characteristic of chordates that most vertebrates keep throughout their lives.

All chordata share four structural characteristics at some point in their life cycle: a dorsal hollow nerve cord, a notochord, gill slits, and a post-anal tail. In most chordata, including humans, these characteristics are present only during early stages of embryonic development.

Human Evolution

Hominids evolved between 5 to 8 million years ago. Fossil records provided evidence of this development and date from about 4.4 million years ago. There were about nine different hominid species.

A species, *homo habilis* probably existed between 2.4 and 1.5 million years ago, and could be a direct ancestor of the modern *homo sapiens. Homo habilis* eventually evolved into *homo erectus,* with a larger brain. The oldest *H. erectus* fossil is about 1.8 million years old. Research indicates that *H. erectus* built fires, wore animal skins, and made some stone tools.

Evidence of the first *Homo sapiens* appeared between 200,000 to 300,000 years ago.

Neanderthals probably lived at the same time as modern humans, but are differentiated by their larger, heavier body size, a longer skull with a bulge at the back, and thick, heavy bones.

Humans could have evolved from *H. erectus* populations in various parts of the world, or from ancient populations in Africa before they dispersed to other areas. Regardless of which theory may be true, the mitochondrial DNA in today's population is nearly uniform. Extracting DNA from Neanderthal and other extinct human fossils might help to establish similarities between ancient and modern humans (See Figure 9.1).

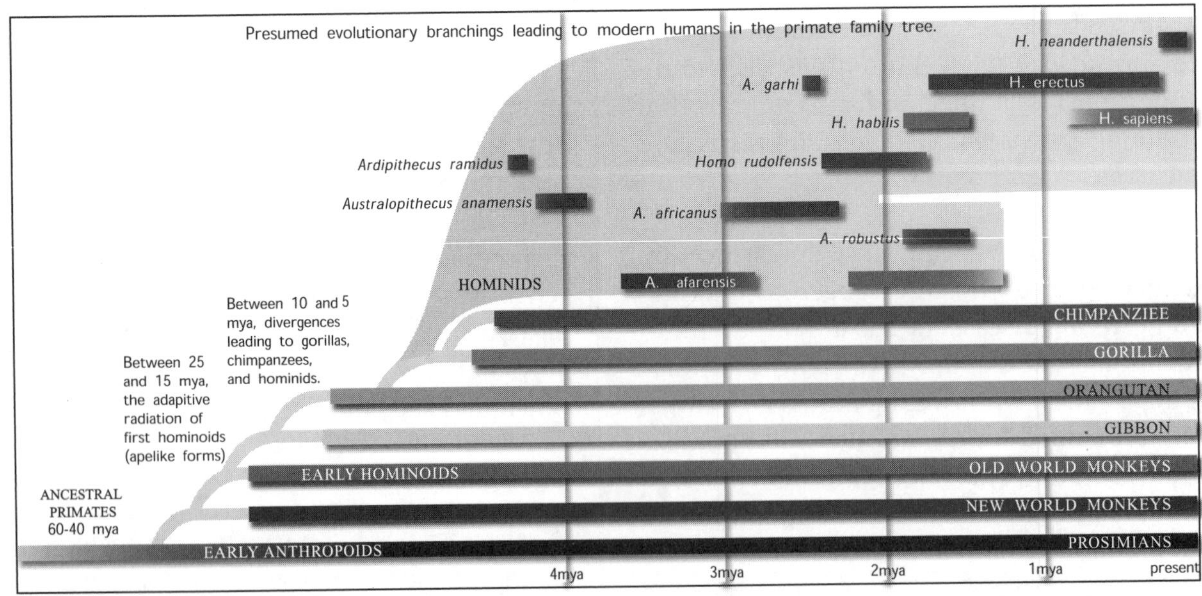

Figure 9.1—Evolution Path of Humans

SUMMARY

- Archaebacteria and eubacteria are important in both the balance of life and in the development of infectious diseases. Protists are both unicellular and multicellular eukaryotes, and include protozoa, algae, and molds.
- Fungi, once classified as part of the plant kingdom, are now in their own group because the cell walls of plants are made of cellulose, while that of fungi is chitin. Fungi cannot make their own food.
- Plants are not only an important food source but are critical for the production of oxygen, which is essential in any environment that can sustain humans and other animals.
- Animals, including invertebrates and vertebrates, illustrate the diversity of life.

ANIMAL FORM AND HOMEOSTASIS

KEY TERMS

tissue	microfilaments	
metabolic rate	endotherm	ectotherm

By now, it is obvious that organization exists on many levels in science. Think life, think evolution, think structure and function, and think biology. We have already seen the relationship between the structure and functions of cells. Now it is time to widen our view to investigate the tissues. In most animals, especially humans and other vertebrates, there are four main categories of tissues: epithelial tissue, connective tissue, muscle tissue, and nervous tissue. In this chapter we will explore the structure of these various tissues and learn their roles in the complex organization of living things.

TYPES OF CELLS AND TISSUES

Epithelial Tissue

Epithelial tissue is a versatile tissue that forms from sheets of tightly packed cells, covers the outside of the body, and lines organs and body cavities. The free surface of this tissue is exposed either to air or fluid. The base of the cells is attached to a basement membrane (a dense layer of extracellular material). The cells are closely joined and may act as a barrier against injury, microbial invasion, or fluid loss (See Figure 10.1).

Biologists categorize epithelial tissue by the number of layers and the shape of the free surface of the cells. Epithelium may be simple (one layer of cells), stratified (multiple tiers of cells), or pseudostratified (one layer that appears multiple because the cells vary in length). The cell shapes are squamous, cuboidal, or columnar. These cells may be specialized for absorption or secretion of chemical solutions. For example, the mucus membrane lining of the mouth and nasal passageways secret mucus, which moistens and lubricates the surface.

Simple squamous epithelium tissue is thin and leaky. The cells function in the exchange of material by diffusion and line blood vessels and the air sacs of the lungs. Stratified squamous tissue regenerates rapidly near the basement membrane. New cells are pushed to the free surface as replacements for the cells that are continually sloughed off. Stratified squamous tissue is on surfaces that are subject to abrasion. The outer skin, for example, is composed of this tissue.

The cells of columnar epithelia are rather like a cytoplasm filled water balloon. You'll find them where secretion or active absorption of substances is an important function, such as in the intestines, where it secretes digestive juices. The small intestine is the major site for absorption of nutrients. Stratified columnar epithelia line the inner surface of the urinary bladder. Pseudostratified ciliated columnar epithelia line the nasal passages of many vertebrates. Simple cuboidal epithelium is specialized for secretion and does its job in

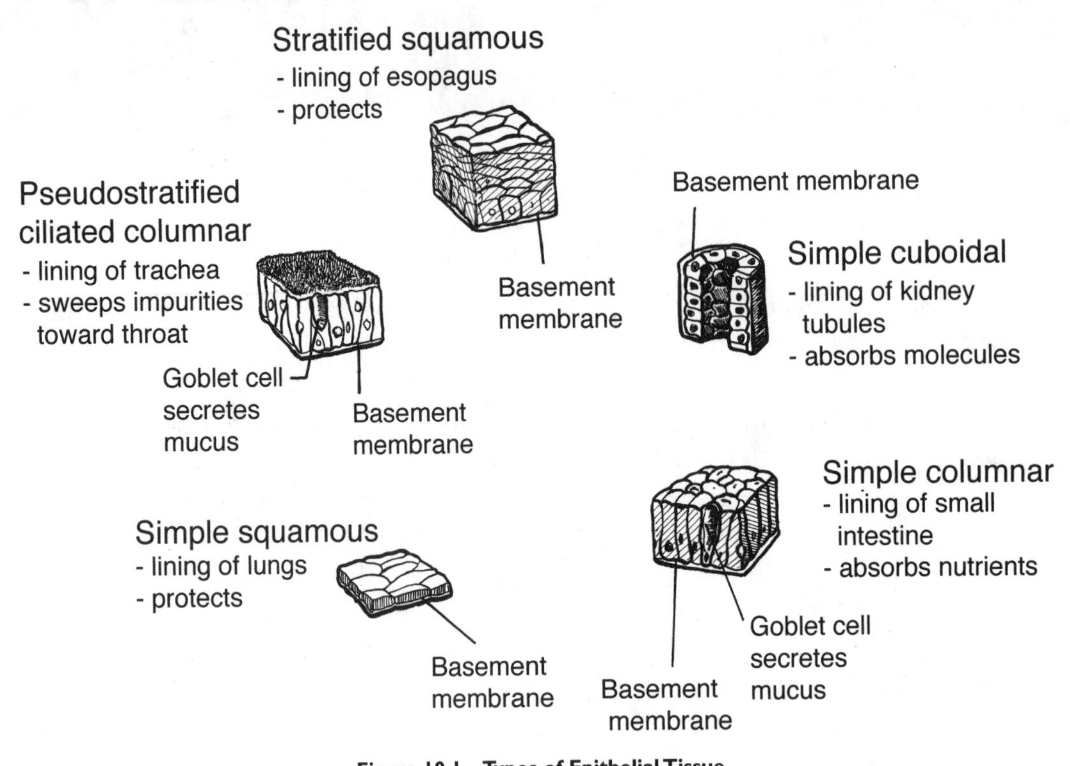

Figure 10.1—Types of Epithelial Tissue

the kidney tubules, the thyroid gland, and the salivary glands.

Connective Tissue

Connective tissue binds and supports other tissues. It has a sparse cell population scattered throughout an extensive extracellular matrix. The matrix is a web of fibers embedded in a substance with the consistency of soft-set gelatin. Three types of fibers make up the various types of connective tissue (See Figure 10.2).

- **Collagenous fibers** are bundles of fiber containing three collagen fibers each. These fibers are strong and resist stretching. You can see collagen bundles by looking at the parallel lines on the palms of your hands.

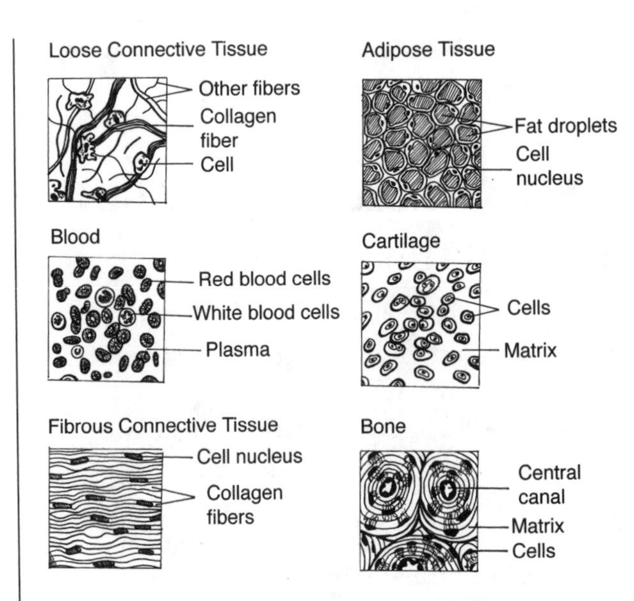

Figure 10.2—Types of Connective Tissue

- **Elastic fibers** are long threads of the protein, elastin. If stretched, this tissue can return to its original shape.
- **Reticular fibers** are branched and tightly woven. They join connective tissue to neighboring tissues.

Connective tissue falls into six categories: loose connective tissue, adipose tissue, fibrous connective tissue, cartilage, bone, and blood.

Loose connective tissue holds organs in place and attaches the epithelium to underlying tissues. It contains two types of cells. The fibroblasts secrete protein of the extracellular fibers. The macrophages serve as the attack dogs of the body's immune defense. Loose connective tissue contains all three fiber types—collagen, elastin, and reticular fibers. Adipose tissue is loose connective tissue that is specialized to store fat in adipose cells distributed throughout its matrix. Each adipose cell stores one fat droplet, which varies in size. The stored fat insulates the body and is used for fuel when needed.

Large numbers of collagenous fibers in parallel bundles make fibrous connective tissue very dense. This gives it the great tensile strength needed for tendons (to attach muscles to bone) and ligaments (to attach bones together at joints). To get an idea of its form, reach down and slide your hand up the back of your ankle along your Achilles tendon.

Cartilage is a strong and flexible connective issue found in the skeleton of all vertebrate embryos. Most vertebrates convert the cartilage to bone, but they retain cartilage in the nose, the ears, the trachea, and the intervertebral discs, where flexibility is a plus. Sharks retain a cartilaginous skeleton throughout their life. Cartilage is composed of collagenous fibers embedded in chondroitin sulfate (a protein-carbohydrate).

Bone is a mineralized connective tissue. Osteoblasts (bone-forming cells) deposit a matrix of collagen and calcium phosphate, which hardens into the mineral hydroxyapatite. Even though bone is hard, it is not brittle; nor is it completely solid. Blood vessels and nerve cells occupy slender canals in the bone tissue, called Haversian canals.

Blood is a liquid extracellular matrix of plasma containing water, salts, and proteins. The cellular component of blood contains red blood cells for the transport of oxygen, the white blood cells of the immune system, and platelets, which are cell fragments that take part in proper clotting of the blood.

- **Blood cells are made in the red bone marrow of long bones.**

Muscle Tissue

Pound for pound there is more muscle tissue in most animals than any other type of tissue. It is made up of long, excitable cells able to contract (See Figure 10.3). The cytoplasm consists of parallel bundles of microfilaments made of the contractile proteins, actin and myosin. Three types of muscles occur in vertebrates.

- **Skeletal muscle** is multinucleated and is usually attached to bones by tendons. Contraction is voluntary and the muscle appears striated under the microscope.
- **Smooth muscle** is found in the walls of internal organs and arteries. The spindle-shaped, uninucleated cells contract involuntarily.

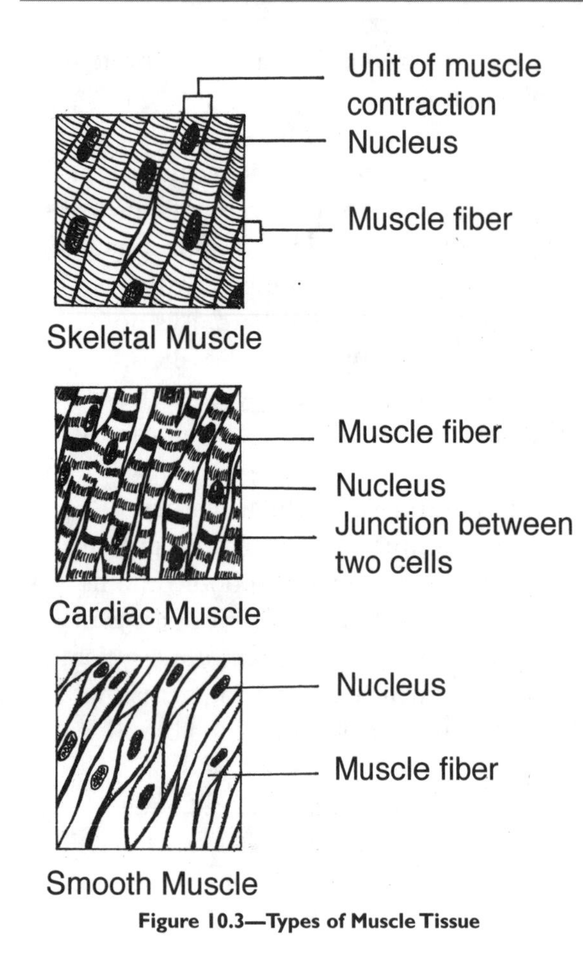

Figure 10.3—Types of Muscle Tissue

- **Cardiac muscle** is only in the wall of the heart. The cells are striated, uninucleated, and joined by intercalated disks. Contraction of cardiac muscle is involuntary.

Nervous Tissue

Nervous tissue is designed to sense stimuli. It transmits signals from one part of the organism to another. The neuron (nerve cell) conducts impulses or bioelectric signals.

ORGANS AND ORGAN SYSTEMS

Tissues can work together to perform a particular task. An organ is a body structure that has definite form and function and consists of more than one type of tissue. Let the human stomach serve as an example. Tissues in the stomach are layered. The mucosa, an epithelial layer, lines the lumen, the space inside. The submucosa is a matrix of connective tissue that contains blood vessels and nerves. The muscularis contains an inner layer of circular muscles and an outer layer of longitudinal muscles. Finally, the stomach has a thin layer of connective tissue and epithelial tissue called the serosa. The cooperative tissue structure of the stomach is perfect for its role in the digestive system (Chapter 13).

An organ system is a team of two or more organs that are interacting chemically and/or physically in a common task. For example, the stomach is part of the digestive system. The digestive system is a continuous tube beginning with the mouth and ending at the rectum. In between are the esophagus, the stomach, the small intestine, and the large intestine. The liver, the gallbladder, and the pancreas, are accessory organs that contribute to the process of digestion.

Bioenergetics

It takes energy to run organ systems, so it is important that energy intake and loss be balanced. Bioenergetics is the study of the dynamic balance of energy in an organism. It gives clues about how an organism adapts to its environment.

Metabolic rate is the total amount of energy an animal uses per unit of time. Metabolic rates are influenced by a variety of factors: age, sex, size, body temperature, environmental temperature, food quality and quantity, activity, amount of available oxygen, hormone balance, and the time of day. Minimal rates support basic life functions and maximal rates of metabolism are displayed during peak activity of the organism.

> Endotherms (birds and mammals) generate their own body heat metabolically. Ectotherms (amphibians and reptiles) acquire most of their heat from the environment. Their body temperature and metabolic rate changes with the environmental temperature.

How do organisms get the fuel needed for the production of energy? Animals simply eat. During the processes of digestion and absorption, food is enzymatically broken down and body cells absorb the small, energy-containing molecules. Cellular respiration harvests the chemical energy from the food molecules. Most of the energy is stored in ATP. It is this molecule that powers the work of the cells. After the needs of staying alive are met, leftover chemical energy and carbon skeletons from food molecules can be used in biosynthesis.

Homeostasis

To stay alive, the cells of an organism need proper nutrients and oxygen, and metabolic wastes must be removed. With this goal in mind, the body size and shape of an organism affect its interactions with its environment. Those interactions can be very different for different species depending on exposure to the environment. For example, the entire single-celled ameba is in contact with its "outside" environment. Every cell of the bilayered hydra is in contact with its environment. More complex organisms are not entirely exposed in the same manner. Even so, the component parts of any organism, small or large, must work together to maintain the stable fluid environment that all of its cells require. This is the concept of homeostasis. It is absolutely central to understanding the structure and function of organisms.

Humans have developed ways to make the home environment liveable. We set the thermostat to sense temperature. We have telephones for communication, and appliances to do some of the heavy chores. Homeostatic control mechanisms help maintain physical and chemical aspects of the body's internal environment within ranges that are most favorable for cell function. How is this state maintained? In humans, there are three components comprising homeostatic control: sensory receptors, integrators, and effectors.

- **Sensory receptors** can detect a stimulus (a change in the environment).
- An **integrator** can send signals regarding stimulus (the brain).
- **Effectors** carry out the response (muscles and/or glands).

Feedback mechanisms are among the controls that operate to keep chemical and physical aspects of the body within tolerable ranges. A positive feedback amplifies positive changes. A variable in the environment triggers the mechanism. In human childbirth, for example, the pressure of the infant's head against sensors in the uterus increases the strength and frequency of the uterine contractions. The brain interprets this increased pressure and increases the production of the hormone, oxytocin, which further increases the uterine contractions.

In a negative feedback mechanism, some activity alters a condition in the internal environment, which triggers a response that reverses the altered condition. Examples of negative feedback abound. If your cells have too much ATP, this substance itself inhibits one of the early enzymes employed in the synthesis of ATP.

SUMMARY

- Tissues are groups of cells with a common structure and function. Epithelial tissue covers and lines the body. Connective tissue binds and supports other tissues. Muscle tissue is involved with movement. Nervous tissue forms a communication network.
- Multiple tissues adapted as a group to perform specific functions form structures called organs. The human body is a cooperative of organ systems, which are interdependent upon one another.
- Bioenergetics gives clues about how an organism adapts to its environment. An animal's body must allow exchange with its environment. A single-celled organism can carry on this exchange by diffusion. More complex organisms contain component parts that work together to allow every cell to interact with its environment.
- Multicellular organisms must regulate their internal environment. Homeostasis is dependent upon positive or negative feedback mechanisms.

CHAPTER 11

INTEGUMENTARY SYSTEM

KEY TERMS

subcutaneous	follicles	melanin
keratinocytes	sebum	epithelium

The skin and its derivatives—hair, nails, glands, and nerve endings—make up the integumentary system. The skin consists of different tissues that are joined to perform specific functions. It is the largest organ of the body in surface area and weight. Of all of the body's organs, none is more easily inspected—or more exposed to infection, disease, and injury—than the skin.

SKIN

The skin is composed of two principal parts. The part you see and touch is the epidermis, the outer, thinner portion composed of layers of epithelium. The dermis is the inner, thicker layer of connective tissue. The dermis is attached by fibers to a layer called the subcutaneous layer, which isn't considered part of the skin. The subcutaneous layer serves largely as a storage depot for fat and contains large blood vessels that supply the skin. The subcutaneous layer is attached to the underlying tissues (See Figure 11.1).

Epidermis

The epidermis is composed of layers of stratified squamous epithelium. It contains four principal types of cells:

- keratinocytes
- melanocytes
- Langerhans cells
- Merkel cells

Keratinocytes make up ninety percent of the epidermal cells and they produce the protein keratin, which is a tough fibrous protein that helps protect the skin and underlying tissues from heat, microbes, and chemicals. Keratinocytes also release a waterproofing sealant for the skin. Eight percent of the epidermal cells are melanocytes. They produce the pigment melanin, which contributes to skin color and absorb ultraviolet light. Once inside the cells, the melanin granules cluster to form a protective veil over the nucleus of the cell on the side of the skin surface. In this way, they shield the genetic material (DNA) from damaging UV light. Langerhans cells make up a small proportion of the epidermal cells and participate in immune responses mounted against microorganisms. Merkel cells

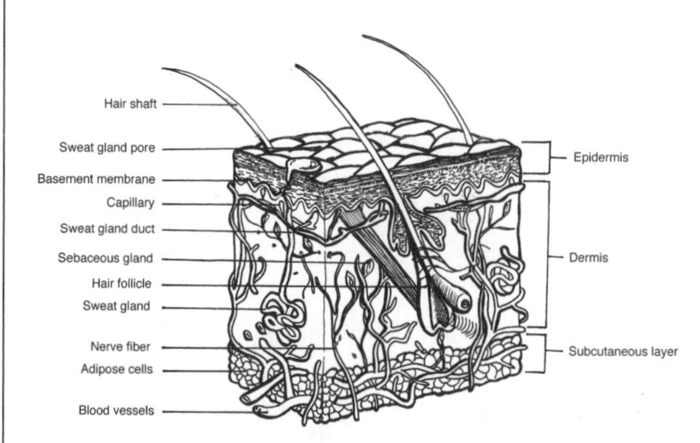

Figure 11.1—Structure of Human Skin

are the least numerous of the epidermal cells and are located in the deepest layer of the epidermis where they are in contact with nerve cells and function in the sensation of touch.

The epidermis is four or five layers thick, depending upon its location. The deepest level is a single layer of stem cells capable of continued cell division, and cells sensitive to touch. Some of these produce keratinocytes that keep moving up in the skin. While these cells are moving to the skin's surface, melanin is taken in by the cells, they lose their nucleus and other organelles, lose the ability to carry on vital metabolic reactions, and eventually die. Some of the stem cells move down in the skin and become the oil and sweat glands and hair follicles.

The most superficial layer of the epidermis, stratum corneum, is composed of about thirty rows of flat, dead cells, completely filled with keratin. This layer is shed continuously and replaced by cells from the lower layers. This serves as an effective barrier against light and heat waves, bacteria and many chemicals. It takes about two to four weeks for the epidermal cells to make their migration from the lowest layer to the skin's surface.

- **The epidermis is approximately one tenth of a millimeter thick, except in areas of high friction (palms and soles), which can be one to two millimeters thick due to an additional epidermal layer.**

Dermis

If you had a childhood mishap and ended up with badly skinned knees, you might have seen your dermis first hand. The dermis is connective tissue with collagen and elastic fibers and has blood vessels, nerves, glands and hair follicles embedded in it. It is very thick in the palms and soles and very thin in the eyelids, penis, and scrotum. The combination of collagen and elastic fibers in the dermis provides the skin with strength, extensibility (ability to stretch), and elasticity (ability to return to its original shape after stretching). Stretching is seen during pregnancy, obesity, and edema. Nerve endings sensitive to cold are found in and just below the dermis. Nerve endings sensitive to heat are located in the middle and outer dermis.

- **Tears in the dermis (striae) can occur during extreme stretching. Striae are visible as red or silvery white streaks on the skin surface.**

Skin Color

Melanin, carotene, and hemoglobin are the three main pigments responsible for skin color. Melanin is a brown-black pigment found mostly in the epidermis and causes variations in skin tone from pale yellow to black. Since the number of melanocytes is about the same for all races, differences in skin color are due mainly to the amount of pigment the melanocytes produce and disperse to keratinocytes. Melanocytes are most numerous in the mucous membranes, penis, nipples of the breasts and the area around the breasts, face, and limbs. Freckles represent melanin-filled cells clustered in patches.

- **As one ages liver spots may form. They are flat skin patches that accumulate melanin.**

In people of Asian ancestry, carotene is found in the outer layer of the epidermis and the fatty areas of the epidermis and subcutaneous layer. It is the combination of carotene and melanin that account for the yellowish hue of the skin.

In Caucasians, the epidermis is translucent because little melanin is present. Their skin may appear pink depending upon the amount of blood moving through the capillaries in the dermis.

Albinism is an inherited inability of an individual of any race to produce melanin. Most albinos have melanocytes, but are unable to synthesize an enzyme (tyrosinase) necessary to produce melanin. Melanin will be absent in their skin, hair, and eyes, thus, their white appearance. In another condition, called vitiligo, the partial or complete loss of melanocytes from patches of skin produces irregular white spots.

Hair

Humans probably fuss more about hair than any other skin derivative. We cut it, bleach it, dye it, style it, shave it, and fret over having it in some places and losing it in others. Hairs are simply growths of the epidermis variously distributed over the body, with the exception of the palms and soles. Genetics and hormonal influences largely determine the thickness and pattern of distribution of hair. In adults, the greatest distribution of hair is most often seen across the scalp, eyebrows, armpits, and around the external genitalia.

- **Normal hair loss is about seventy to one hundred hairs per day.**

The main function of hair is protection. It guards the scalp from injury and the sun's harmful rays. Eyebrows and eyelashes protect the eyes from foreign particles. Hairs in the nostrils and external ear canal have a similar function. Hair also aids in sensing light touch. Touch receptors associated with hair follicles are activated when a hair moves even slightly. This is important in many animals, such as cats and mice who use the whiskers of the face to move in dark or enclosed spaces.

Each hair is composed of columns of dead, keratinized cells welded together. The shaft is the superficial portion of the hair that projects from the surface of the skin. The shaft of straight hair is round in cross section while that of curly hair is oval. The root is the part of the hair that penetrates into the dermis and sometimes the subcutaneous layer. Surrounding the root is the hair follicle. The base of the hair follicle is shaped like an onion and is called the bulb. This region houses the blood vessels, and the cells of the matrix which are responsible for the growth of existing cells and the production of new hairs by cell division when older hairs are shed. Around each hair follicle are nerve endings that are sensitive to touch. Smooth muscle (arrector pili) is attached to the hair follicles. When these muscles contract under conditions of fright, cold, etc., hair is pulled into an upright position. Now you know what is happening when you get so scared that the hair stands up on the back of your neck. Another expression of this hair-raising effect is "goosebumps."

Hair color is due primarily to the pigment melanin, which is synthesized by the melanocytes scattered in the matrix of the bulb and is then passed into the hair. Dark colored hair is mostly due to true melanin. Blond and red hair are due to variants of melanin in which there is iron and sulfur. Gray hair is a progressive loss of the enzyme responsible for the formation of melanin. White hair is the result of accumulation of air bubbles in the hair shaft.

GLANDS

The four principal types of glands associated with skin are oil, sweat, and two modified sweat glands, ceruminous and mammary.

Oil Glands

Oil glands vary in size depending where they are located. They are large in the skin of the breasts, face, neck, and upper chest. They are small in most areas of the trunk and the extremities, and absent on the palms and soles. Oil glands are most frequently connected to hair follicles. These glands produce an oily substance (sebum) containing a mixture of fats, cholesterol, proteins, inorganic salts, and pheromones. This oil has several functions. It coats the surface of hair and keeps it from drying and becoming brittle. Sebum prevents excessive evaporation of water from the skin. It keeps the skin soft and pliable and inhibits the growth of certain bacteria.

Sweat Glands

Sweat glands, numbering three to four million, empty their secretions to the skin's surface. They fall into two categories—eccrine and aprocrine—depending on structure, location, and type of secretion. Eccrine glands are simple, coiled, tubular, and more common than apocrine glands. Eccrine glands begin functioning soon after birth. Sweat or perspiration is mainly produced by the eccrine glands, which cover many areas of the body. They are most numerous in the skin of the forehead, palms, and soles, where their density can be as high as 450 per square centimeter. The secretory portion of the sweat gland is located mostly in the deep dermis. The secretory duct projects through the dermis and the epidermis and ends as a pore on the surface of the skin. Sweat is a mixture of water, salts, urea, uric acid, amino acids, sugar, lactic acid, and ascorbic acid. Its main function is to help regulate body temperature by providing a cooling mechanism. As sweat evaporates, large quantities of heat energy leave the body surface. It also plays a small role in eliminating wastes.

Apocrine glands begin functioning at puberty. They are simple, coiled tubular glands located mainly in the subcutaneous layer and their excretory ducts open into the hair follicles. The secretion is more viscous than sweat due to additional lipids and proteins. These glands are located in the armpits, pubic region, and pigmented regions of the breast. Aprocrine glands are stimulated during emotional stress and sexual excitement.

Ceruminous glands are modified sweat glands located in the subcutaneous layer of the external ear. They produce a waxy substance that acts as a sticky barrier preventing the entrance of foreign bodies. Mammary glands are also modified sweat glands.

Nails

Nails are plates of tightly packed, hard, keratinized cells of the epidermis. The nail body is the visible portion of the nail. Most of the nail body is pink due to the blood flowing through underlying capillaries. The nail root is buried in a fold of skin. The epithelium underneath the nail root is called the nail matrix and is composed of cells undergoing mitosis to produce growth of the nail. Nails help us grasp and manipulate small objects, provide protection against trauma to the tips of our fingers, and, help us to scratch when we itch. Nails can grow at various speeds, but average about one millimeter per week.

- **The longer the finger, the faster the nail grows.**

Skin Functions

One important function of the skin is protection. Skin covers the body and provides a physical barrier that protects underlying tissues from physical and bacterial invasion, dehydration, and ultraviolet light. Certain cells of the epidermis play a role in the immune system, warding off invaders. Hair and nails provide protection, also.

Skin serves many other purposes as well. It plays a large role in the regulation of body temperature. Under conditions of high environmental temperature or strenuous exercise, glands produce sweat and the evaporation of the sweat provides a cooling mechanism for the body. Under conditions of low environmental temperature, the production of sweat decreases. Sweat also contributes to excretion. Along with heat and some water, it removes a small amount of salts, and several organic compounds.

The dermis contains an extensive network of blood vessels that carry up to ten percent of the total blood flow in a resting adult. For this reason, the skin acts as a blood reservoir. Nerve endings that detect temperature, touch, pressure, and pain are located in the skin. Synthesis of vitamin D begins with the activation of a precursor in the skin by UV. This compound is modified and becomes calcitriol, the most active form of vitamin D. Calcitriol aids in the absorption of calcium in foods from the digestive tract into the blood.

Skin and Aging

Notwithstanding creams, lotions, facial treatments and cosmetic surgery, time eventually catches up with the skin. Changes may be noticeable in a person by their late forties (See Figure 11.2). Collagen fibers decrease in number, stiffen, break apart, and degenerate into a shapeless, matted tangle. Elastic fibers lose some of their elasticity, thicken into clumps, and fray. As a result, the skin forms crevices and furrows, which we see as wrinkles. The cells that produce collagen and elastic fibers decrease in number.

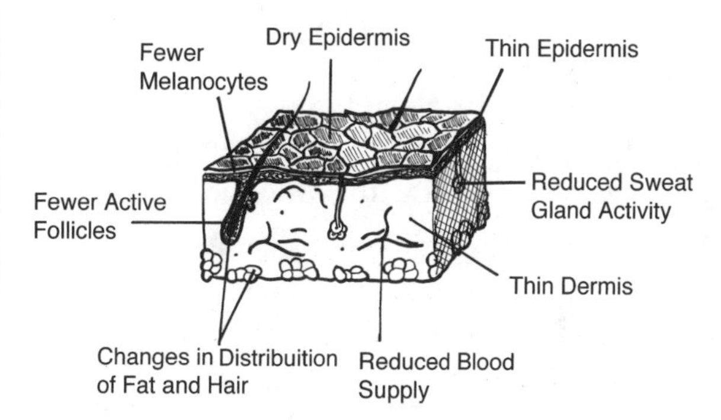

Figure 11.2 — Changes in the Skin During Aging

Oil glands decrease in size, and the result is dry and cracked skin that is more susceptible to infection. Langerhans cells, too, are fewer in number, and the immune responsiveness is slower in older skin. Macrophages become less efficient phagocytes. Aged skin heals slower than younger skin and is more susceptible to pathological conditions such as itching, pressure sores, shingles, and skin cancer. Older skin is thinner than younger skin, especially, the dermis. The migration of cells from the bottom layer to the top slows, and subcutaneous fat is lost. An increase in size of some of the melanocytes results in liver spots. A decrease in the number of functioning melanocytes is responsible for gray hair. Finally, hair and nails grow more slowly.

- **The production of sweat diminishes with age so seniors are more prone to heat stroke.**

SKIN DISORDERS

Acne

Seniors may have to deal with aging skin, but teens and young adults have their own problems with acne. Acne is an inflammation of oil glands that usually begins at puberty when the oil glands grow in size and increase the production of sebum. Steroid hormones play a large role in stimulating the oil glands. Bacteria can colonize sebaceous follicles and cause infections. Antibiotics need to be taken if infection is present. Topical vitamin A ointments have met with some success.

Psoriasis

Psoriasis is a chronic, noncontagious relapsing skin condition characterized by red skin elevations covered with scales. It usually involves the scalp, elbows and knees, the back, and the buttocks. With psoriasis there is an abnormally high rate of mitosis in the epidermal cells. Various causes have been implicated including viruses and defects in the immune system. Steroids, compounds similar in structure to vitamin A, and natural sunlight are some of the therapies for this condition.

Skin Cancer

Excessive exposure to the sun can cause skin cancer, a disease that is on the rise. There are three common forms of skin cancer. The tumors of basal cell carcinomas arise from the lowest layer of the epidermis and account for about seventy-eight percent of skin cancer. They rarely spread. Most squamous cell carcinomas arise from preexisting lesions of skin tissue damaged from the sun. This accounts for twenty percent of skin cancers. Both of these types of skin cancer can be removed surgically and there is a high cure rate. Malignant melanomas arise from melanocytes and account for the remaining two percent of skin cancers. They are the most severe because they metastasize (spread) rapidly and can kill a person within months of diagnosis. The key to successful treatment of malignant melanoma is early detection. People at greater risk (those with light hair, skin, eyes, and those who spend a great deal of time in the sun) should have moles checked frequently for change in color, size and texture.

SUMMARY

- The integumentary system consists of the skin and its accessory structures—hair, glands, and nails.
- The skin is the largest organ of the body in surface area and mass. The two principal regions of the skin are the epidermis, consisting of several layers, and the dermis. The bottom-most epidermal layer is composed of rapidly dividing cells necessary to replace the dead keratinized cells of the outermost layer. Keratinocytes, melanocytes, Langerhans cells, and Merkel cells comprise the epidermis.
- The dermis is below the epidermis and is composed of connective and elastic tissue. The color of the skin is due to melanin, carotene, and hemoglobin.
- Hair is composed of dead keratinized cells and offers limited protection against the

sun, heat loss, and entry of foreign objects into the eyes, ears, and nose. Hair also functions in the sensation of light touch.

- There are four types of glands: oil, sweat, ceruminous, and mammary glands. Oil glands are normally connected to the hair follicles. The sebum produced by these glands moistens hair and waterproofs the skin. Eccrine and apocrine glands are the two main types of sweat glands. Eccrine glands have an extensive body distribution, while the apocrine are localized in the armpits, pubic region, and the nipple region of the breasts.
- The skin provides physical, chemical, and biological barriers that protect the body. Skin functions include body temperature regulation, protection, sensation of pain, heat, cold, and touch, excretion, and synthesis of vitamin D.
- Aging of the skin begins during the late forties. Among the effects of aging are wrinkling, loss of subcutaneous fat, loss of elasticity, decrease in size of the oil glands, less sweat production, and decrease in the number of melanocytes and Langerhans cells.
- During puberty, oil glands can become clogged and acne will result. Psoriasis is a noncontagious condition in which the skin becomes very scaly, due to overproduction of the epidermal cells.

CHAPTER 12 MUSCULOSKELETAL SYSTEM

KEY TERMS

myosin	actin	sarcomere
steroid	intercalated disk	

While we sleep, we rely on our body to pump blood through our arteries and veins, to inhale and exhale, and to change position in bed. When we wake up, we open our eyes, turn off the alarm, jump out of bed, and prepare breakfast. All of these routine functions, crucial to our daily activity, are dependent on the musculoskeletal system.

MUSCLES—STRUCTURE AND FUNCTION

Muscle tissue has four special properties that enable it to function.

- **Electrical excitability** is the ability to receive and respond to stimuli (changes in the environment strong enough to initiate a nerve impulse).
- **Contractility** is the ability to shorten and thicken when a sufficient stimulus is received.
- **Extensibility** is the ability to stretch (extend).
- **Elasticity** is the ability to return to its original shape after contraction or extension.

Through contraction, muscles perform three important functions. When muscles are integrated with bones and joints, they contribute to body movements like walking, running, grasping, nodding the head, etc. Other muscles control movement without any attachment to bones or joints. The heartbeat, contraction of the gallbladder, and contractions of the digestive tract are made possible by muscle activity. Even standing still is hard work. Contraction of skeletal muscles keeps the body in stationary positions and, therefore, has a role in maintaining posture. A byproduct of muscle contraction is heat. This heat helps to maintain the body temperature. Involuntary contractions of skeletal muscles, known as shivering, can increase the rate of heat production several fold.

There are three types of muscles—skeletal, smooth, and cardiac. (See Chapter 10, Figure 10.3) They differ in structure, function, and location.

Skeletal Muscle

Skeletal muscle accounts for between forty and fifty percent of the weight of an individual (See Figure 12.1). Most skeletal muscles are attached to bones. The muscle tapers at its end forming a dense connective cord (tendon) which links the muscle to the bone. Most of the six hundred skeletal muscles mobilize parts of the skeletal system. Some are not attached to bone and these include the muscles involved in eyelid closing and those that control the flow of urine from the urinary bladder. Some skeletal muscles are categorized as flexor or extensor. When a flexor muscle

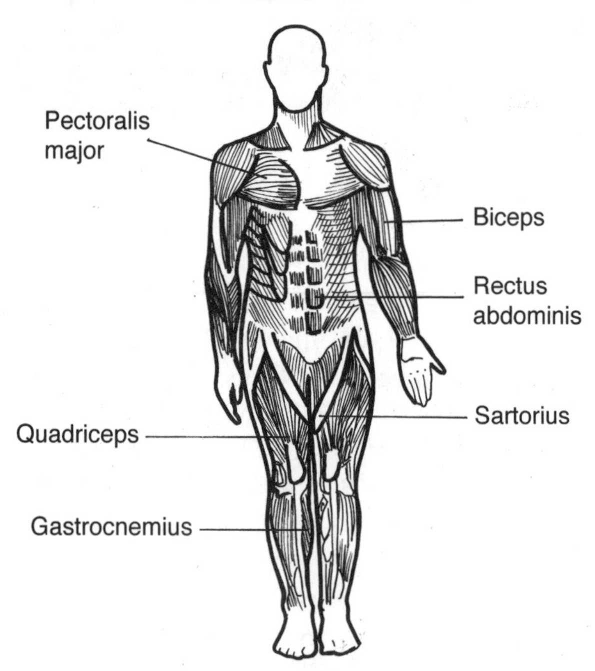

Figure 12.1—Skeletal Muscle

(e.g., the biceps) contracts, it causes a joint to bend. When an extensor muscle (e.g., the triceps) contracts, the joint straightens. Muscles shorten and pull. The biceps/triceps relationship is referred to as an antagonistic muscle pair because one muscle reverses the effects of the other (See Figure 12.2). Neural orders that command a muscle to contract will also inhibit the antagonistic muscle. Skeletal muscles are known for the speed of their contraction.

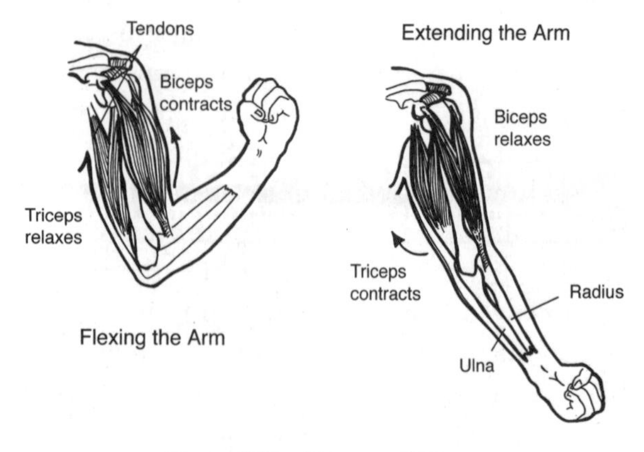

Figure 12.2—Biceps and Triceps

The muscle cell is cylindrical in shape and can be up to one hundred micrometers in length. It is more aptly referred to as a muscle fiber because it runs the length of the entire muscle. Because of its large size, there are thousands of nuclei per cell. The muscle consists of bundles of muscle fibers. Each fiber is covered by a continuous plasma membrane and packed with myofibrils (fine longitudinal fibers). Each of the fibers consists of a linear array of contractile units called sarcomeres (See Figure 12.3). The sarcomere is composed of two types of contractile units, thin (actin) and thick (myosin). This gives the muscle fiber a characteristic banding pattern when viewed through a microscope. Each bundle is enclosed in a connective tissue sheath. Skeletal muscle has a striated, multinucleated, unbranched appearance under the microscope.

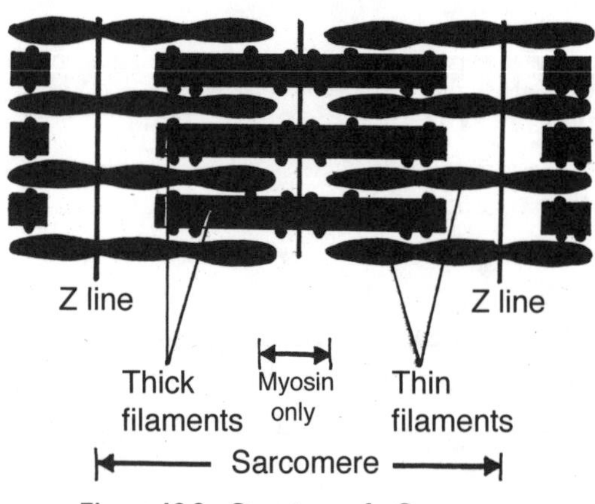

Figure 12.3—Structure of a Sarcomere

- **The contraction of skeletal muscle is voluntary and under nervous control.**

Smooth Muscle

Smooth muscles often participate in processes related to maintaining the internal environment of the body. They are present in the walls of hollow internal structures, including blood vessels and the lining of the gastrointestinal tract. Smooth muscle is also attached to hair shafts. The fibers of smooth muscles are smaller than skeletal muscles. They are spindle-shaped with a single, oval, centrally-located nucleus. The myofilaments have no regular pattern of organization, so you won't see striations under the microscope. Because of its structure, smooth muscle takes longer than skeletal muscle to trigger a contraction and longer for the contraction to subside. Their contraction is involuntary.

- **All muscle cells demonstrate a unique property—once formed they never undergo cell division. They can get longer and wider, but they never undergo mitosis.**

Cardiac Muscle

Cardiac muscle is the principal constituent of the heart wall. It has a unique combination of properties. Like skeletal muscle, cardiac muscle is striated. Like smooth muscle, cardiac muscle is under involuntary control and has a single, centrally-located nucleus. Cardiac muscle tissue needs a constant supply of oxygen. The cytoplasm of cardiac muscle is more abundant than in skeletal muscle, and the cell's mitochondria are larger and more plentiful. Unlike other muscle tissues, cardiac muscle cells are lined up end to end and joined to each other by a dense band (intercalated disk). These disks strengthen cardiac muscle tissue and help in the electrical impulse conduction from one muscle fiber to another. When a single fiber is stimulated, all are stimulated, and they contract in a synchronous fashion. A specialized area called the sinoatrial node (pacemaker) is responsible for the stimulation of contraction of cardiac muscle.

- **Cardiac muscle remains contracted ten to fifteen times longer than skeletal muscle due to a prolonged delivery of calcium ions. Cardiac muscle can relax between beats.**

MUSCLE CONTRACTION

Where does the energy for muscle contraction come from? Muscles can synthesize ATP. However, the storage of ATP in muscles may be depleted in seconds. A high-energy compound (phosphocreatine) in muscle fibers converts ADP to ATP (see page 25), and provides additional energy for muscle contraction. Once that is used up, glycogen in the skeletal muscle and liver is broken down to glucose, which will provide energy for several minutes of activity. Once the glycogen supply has been exhausted, muscles break down fat to provide the energy to resynthesize ATP. The supply of fats is almost unlimited and is provided by regular meals!

The model describing the contraction of muscles is called the sliding filament mechanism. ATP binds to a myosin head and the head detaches from a binding site on actin. Next, ATP is broken down to ADP plus phosphate with the release of energy. The head gains some of this energy and its position changes. Calcium is necessary for muscle contraction. It acts as a molecular trigger by opening up the binding site on the actin molecule. This makes it possible for the myosin head to bind to the actin. The myosin head bends when ADP and phosphate are released from it. Bending pulls the thin filament toward the center of the sarcomere. The power stroke is the molecular event that actually causes sliding of the thin filaments past the thick filaments. The sliding shortens the sarcomere and the muscle contracts.

Of the total energy released during muscle contraction, only a small amount is used for mechanical work. The production of heat by the contraction of skeletal muscles is an important mechanism for maintaining normal body temperature. As much as eighty-five percent of the energy can be released as heat.

Charlie horse, tic, cramp—by any other name they are abnormal contractions of skeletal muscle. A spasm is a sudden involuntary contraction of a single muscle in a large group of muscles. A cramp is a painful spasmodic contraction. A tremor is a rhythmic, involuntary, purposeless contraction of opposing muscle groups. Fasciculation is an involuntary, brief twitch of a muscle under the skin and occurs irregularly. Fibrillation is similar to fasciculation, but not visible under the skin. A tic is spasmodic twitching made involuntarily by muscles that are under voluntary control. These can occur in the eyelids and facial muscles.

MUSCLE DISORDERS

Fibromyalgia

Fibromyalgia appears between the ages of twenty-five and fifty and is fifteen times more common in females than in males. It affects the fibrous connective tissue components of muscles, tendons, and ligaments. Pain, tenderness, and stiffness of muscles, tendons, and the surrounding soft tissue are present. It is caused or aggravated by physical or mental stress, trauma, exposure to dampness or cold, poor sleep, or a rheumatoid condition. It is more common in the lumbar region (lumbago), neck, chest, and thighs.

Muscular Dystrophy

Muscular dystrophy is a sex-linked inherited muscle-destroying disorder. It affects about one in 3500 males. There is a degeneration of individual muscle fibers and a progressive atrophy of skeletal muscle. Usually voluntary muscles are weakened equally on both sides of the body. The internal muscles are not affected. The condition appears between the ages of three and five. The patient eventually becomes wheelchair bound. A blood test detecting the presence of an enzyme (creatine phosphokinase) in large amounts confirms a diagnosis. There is no cure.

SKELETON—STRUCTURE AND FUNCTION

Structurally, the skeletal system is composed of two types of connective tissue: bone and cartilage. Bone tissue contains a great deal of intercellular substance surrounding widely separated cells, called osteocytes. The intercellular substance of bones contains abundant mineral salts, primarily calcium phosphate and calcium carbonate. As these salts are deposited, the bones become ossified (hardened).

Bone is not a completely solid, homogenous substance (See Figure 12.4). In fact, all bone has some spaces between hard components. The spaces provide channels for blood vessels that supply bone cells with nutrients, and they make bones lighter. Depending upon the size and distribution of the spaces, the regions of a bone may be categorized as spongy or compact. Spongy bone tissue contains many large spaces filled with red marrow. It makes up most of the bones of the short, flat, and irregularly shaped bones, and most of the ends of long bones. In contrast, compact bone is dense tissue with fewer spaces. It is deposited in a layer over the spongy bone tissue. Compact bone tissue provides protection and support and helps the long bones resist the stress of weight placed on them.

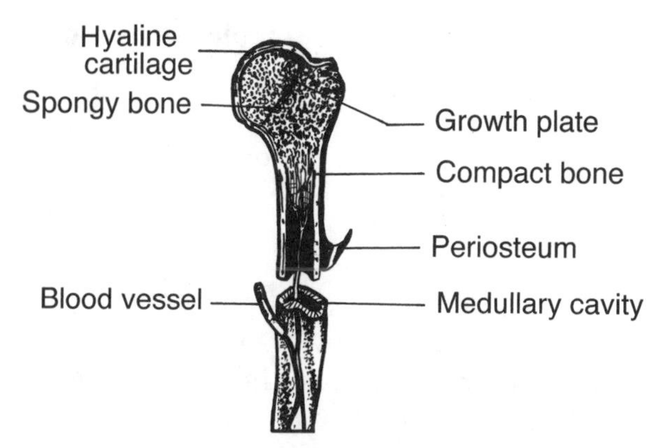

Figure 12.4—Structure of Bone

A main function of the skeletal system is support. The bones are the framework of our body and support the soft tissue. They provide a point of attachment for many muscles, and the skeleton protects internal organs from injury. Cranial bones protect the brain. The vertebrae protect the spinal cord. The ribcage houses the heart and lungs, and the pelvic bones safeguard the internal reproductive organs.

Bones serve as levers to which muscles are attached. When the muscles contract, the bones produce movement. They are a major site of mineral storage, particularly calcium and phosphorus. These minerals can be distributed to other parts of the body upon demand. Red bone marrow in certain bones is capable of producing blood cells. It consists of blood cells in immature stages (stem cells), fat cells, and macrophages. Red marrow produces red blood cells, some white blood cells, and platelets.

- **As you age, some of the marrow becomes yellow due to the storage of triglycerides.**

Bone shares with skin the feature of replacing itself throughout adult life. During remodeling, worn or injured bone is removed and replaced by new tissue. The distal end (furthest from the center) of the femur is replaced about every four months. By contrast, bone in certain areas of the shaft will not be completely replaced during an individual's entire life. This process also allows the bone to serve as the body's storage area for calcium. The blood continually trades off calcium with the bones, removing calcium when it and other tissues are not receiving enough and resupplying the bones with dietary calcium to keep them from losing too much bone mass.

Normal bone growth in the young and bone replacement in the adult depend on several factors. Sufficient quantities of calcium and phosphorus, components of the primary salt that makes up the bone, must be included in the diet. Humans need adequate amounts of vitamins. Vitamin D in particular participates in the absorption of calcium from the digestive tract into the blood, calcium removal from bone, and kidney absorption of calcium that might otherwise be lost in urine.

The sex hormones (steroids) act as a double-edged sword. They aid in the growth of new bone, but they also bring about a degeneration of all the cartilage cells in the epiphyseal plates (responsible for the lengthwise growth of long bones). Because of the sex hormones, the typical adolescent experiences a growth spurt during puberty, when the steroid levels start to increase. The individual then quickly completes the growth process as the epiphyseal cartilage disappears. Premature puberty can actually prevent an individual from reaching an average adult height because of the simultaneous premature degeneration of the plates.

The body must manufacture the proper amounts of various hormones responsible for bone tissue activity:

- **Growth hormone,** secreted by the pituitary gland, is responsible for the general growth of bones.
- **Calcitonin,** produced by the thyroid gland, accelerates calcium absorption by bones.
- **Parathyroid hormone,** secreted by the parathyroid glands, releases calcium from the bones into the blood.

Almost all of the bones of the body may be classified into four principal types on the basis of shape: long, short, flat, and irregular.

- **Long bones** have greater length than width and are slightly curved for strength. A curved bone is structurally designed to absorb the stress of the body at several points so the stress is evenly distributed. If these bones were straight, the weight of the body would be unevenly distributed and the bone would break more easily. Bones of the thighs, lower legs, toes, arms, forearms, and fingers belong in this category.
- **Short bones** are somewhat cube-shaped and nearly equal in length and width. Their texture is spongy, except at the surface, where there is a thin layer of compact bones. Examples of short bones are the wrist and ankle.
- **Flat bones** are generally thin and composed of two or more parallel plates of compact bone enclosing a layer of spongy bone. They afford considerable protection and provide ample area for muscle attachment. Flat bones include the cranial bones, sternum, ribs, and the scapulas.
- **Irregular bones** have complex shapes and cannot be grouped into any of the three categories described. They also vary in the amount of spongy and compact bone. The vertebrae and some of the facial bones are irregular.

There are two additional types of bones, which don't fall into any of the above classifications. Sutural (Wormian) bones are small bones between joints of certain cranial bones. Their number varies greatly from person to person. Sesamoid bones are small bones in tendons where considerable pressure develops, e.g. the wrist. These bones vary in number, also. The kneecap (patella) falls into a category of its own.

The adult human skeleton contains about 206 bones grouped into two principal divisions—the axial skeleton and the appendicular skeleton (See Figure 12.5). There is a considerable difference between the male and the female skeletons. The bones of the male are generally larger and heavier than those of the female. The ends of bones are thicker in relation to the shafts in males. In addition, since certain muscles of the male are larger than those of the female, the points of attachment are larger in the male skeleton.

- **The pelvic girdle of females is designed to allow the birth of children.**

Axial Skeleton

The axial skeleton includes the bones that lie along the longitudinal axis—cranial and facial bones of the skull, auditory ossicles, hyoid, vertebral column, sternum and ribs. This structure assumes the skeleton's protective functions. The skull, which is composed of all of the bones of your head, provides excel-

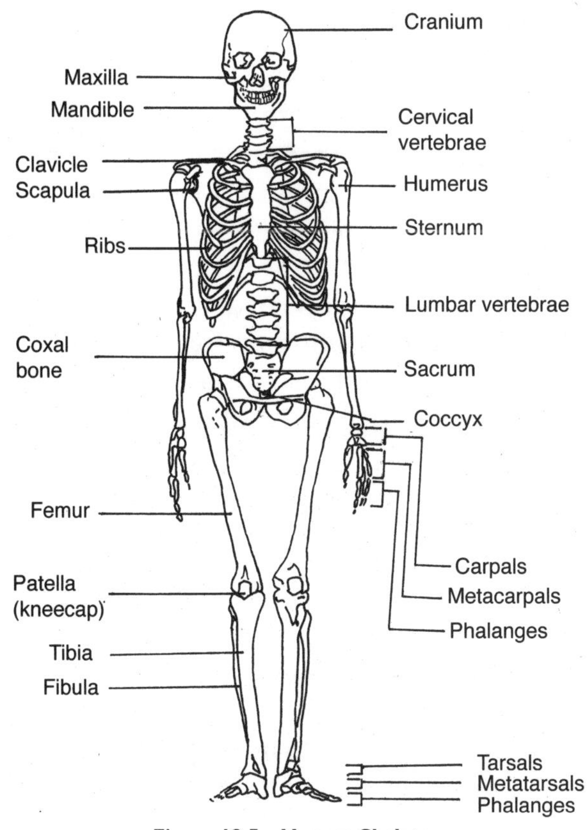

Figure 12.5—Human Skeleton

lent protection to your brain. The hyoid bone is unique because it does not articulate with any other bone. Ligaments and muscles suspend this bone. It is located in the neck between the mandible and the larynx. It supports the tongue and provides attachment for some of its muscles. The hyoid is frequently fractured during strangulation!

- **The auditory ossicles (three in each ear) are the smallest bones in your body.**

The flexible backbone consists of twenty-six vertebrae, which are arranged in a gracefully curved line and are cushioned from one another by disks of cartilage. These intervertebral disks take on a particular importance in humans since walking upright on two legs causes the backbone to bear the entire weight of the upper body. If not for these disks, the vertebrae would grind one another to dust. Sitting on a chair can place even greater stress on the vertebrae, often leading to back pain, a common drawback of a sedentary lifestyle. The vertebral column has four natural curves, which increase its strength, help us to maintain balance in an upright position, absorb shocks from walking and help protect the column from fracture.

Ribs extend from the vertebrae and attach to the sternum on the ventral side of humans. The lower two pairs of ribs remain free at their outer ends. The marrow found in ribs is one of the most prolific producers of red blood cells.

Appendicular Skeleton

The movable limbs attached to the axial skeleton comprise the appendicular skeleton, which forms a system of levers, providing mobility and dexterity. The pectoral girdle holds the arms to the axial skeleton. It consists of two scapulae and two clavicles. The human arm has tremendous dexterity. The human hand combines strength and dexterity.

- **The clavicle is the most frequently broken bone in the body—it can break during birthing.**

The pelvic girdle receives the weight of the upper body from the vertebral column and transmits it either to the leg bones or to the surface on which you are sitting. The pelvis differs in shape between males and females to accommodate birthing. Although the foot is not as dexterous as the human hand, it is able to withstand tremendous forces. The reason for this is the arched shape of the foot bones.

An arch is the most efficient structure for supporting weight. Think of some engineering accomplishments that have utilized the arched shape.

Joints and Ligaments

The site where two bones come together is called a joint. In general, the more mobile the joint is, the weaker it is. Sutural joints join the plates of the skull. They are very strong and virtually immobile after the age of two. In contrast, the shoulder is the most movable joint making it vulnerable to dislocation. The shoulder is a ball and socket joint. A hinge joint allows back and forth movement as demonstrated by the knees and elbows. A pivot joint allows rotation of the forearm at the elbow. The adjoining bones of a joint are held together by strong straps of connective tissue called ligaments. The strength of ligaments is one of the principal mechanical factors that hold bones close together. A sprain is the forcible wrenching or twisting of a joint that stretches or tears its ligaments but does not dislocate the bones.

Cartilage consists of a dense network of collagen fibers and elastic fibers. Hyaline cartilage is the most abundant cartilage and affords flexibility and support, and, at joints, reduces friction and shock. Fibrocartilage combines strength and rigidity and is the strongest of the cartilages. It is the material of the intervertebral disks. Elastic cartilage provides strength and elasticity and maintains the structure of the external ear.

Coordination of the Bones and Muscles

Muscles, bones, and joints function individually, but they must have coordinated activity to perform mechanical work. The body uses bones as levers and joints as fulcrums. The movement of a part of the body is accomplished by a force (generated by muscle contraction) acting on levers (represented by bones) that pivot on a fulcrum (one of the body's joints).

Skeletal Disorders

Fractures

You probably rarely think about the stresses your bones endure in a day of regular activity. That is, until you experience a broken bone, or fracture. Any break in a bone is referred to as a fracture. In a partial fracture, the break across the bone is incomplete, whereas in a complete fracture, the bone is in two pieces. There are several types of fractures.

- **Closed (simple) fracture:** the skin is not broken.
- **Open (compound) fracture:** the bone protrudes through the skin.
- **Greenstick fracture:** only seen in children. One side of the bone is broken while the other side bends.
- **Impacted fracture:** one fragment is firmly driven into another.
- **Stress fracture:** partial fracture resulting from an inability to withstand repeated stress due to a change in training, harder surfaces, longer distances, or greater speeds. These fractures are often seen in runners and joggers.

During a fracture, blood vessels break, blood pours out, coagulates and forms a clot in and around the fracture site. Growth of new bone tissue develops in and around the fractured area within forty-eight hours of the fracture. Dead portions of the fracture are reabsorbed. Compact bone replaces spongy bone on the outside, and, so a bone will always be thicker after a fracture.

Osteoporosis

Osteoporosis is an age-related disorder affecting more women than men and more whites than blacks. It generally appears after menopause. There is a decrease in bone mass due to a decline in the number of bone-producing cells and, therefore, an increased risk of fractures. Osteoporosis affects the whole skeletal system, but mainly the spine, hips, legs, and feet. It is responsible for shrinkage of the backbone, hunched backs, hip fractures, and plenty of pain. Various factors: decreased levels of estrogen, calcium deficiency and malabsorption, vitamin D deficiency, loss of muscle mass, and inactivity can aggravate the condition. Calcium supplements and increased weight-bearing exercise may help the condition. Estrogen replacement therapy has been prescribed in the past for this condition, but has recently come under attack due to side effects.

Vitamin D Deficiency

Children with a vitamin D deficiency can develop rickets. In such a child, there is an inability of the body to transport calcium and phosphorus from the digestive tract to the blood to the bones. Old cartilage does not degenerate, new cartilage is laid down at the ends of the bones, and the bones do not harden. As the child grows and gets heavier, the leg bones bow. Cure and prevention include having a diet with sufficient calcium, phosphorus, and vitamin D.

> Osteomalacia is caused by a vitamin D deficiency in adults. Here there is a demineralization of bone, which leads to bowing of the legs, shortening of the backbone, and flattening of the pelvis bones. Treatment is the same as for rickets.

SUMMARY

- The human body is composed of approximately six hundred muscles. They are described as skeletal, smooth or cardiac muscle depending upon their structure, function, and location.
- Skeletal muscles are usually attached to bones and are responsible for voluntary movement. They are composed of bundles of striated, multinucleated cells. Smooth, involuntary muscle is composed of uninucleated, nonstriated cells. The muscles of your internal organs are composed of these spindle-shaped cells.
- Cardiac muscle, found only in the heart, consists of striated, uninucleated cells that are connected by intercalated disks.
- Muscles contract by the sliding filament mechanism. Besides being responsible for movement, muscles can generate as much as eighty-five percent of the heat that humans need to maintain their body temperature.
- Abnormal muscle contractions of skeletal muscle can cause spasms and tics to occur. Fibromyalgia is a muscle disorder of vague causes and is difficult to treat. Muscular

dystrophy is a genetic deterioration of skeletal muscles.

- The skeletal system functions in support and protection. In addition, bone is a major site for the storage of calcium and phosphorus. The red bone marrow is the site of blood cell manufacture.
- Most bones can be classified according to their shape—long, short, flat, and irregular. The two principal divisions of the skeletal system are the axial skeleton and the appendicular skeleton.
- The axial skeleton includes the bones that lie along the longitudinal axis, while the movable limbs attached to the axial skeleton comprise the appendicular system. Differences are distinguishable between the male and female adult skeleton.
- Joints, ligaments, and cartilage join the muscles and the bones to form coordinated body activities.
- Skeletal disorders include fractures of bones, vitamin D deficiency, and osteoporosis.

CHAPTER 13

DIGESTIVE SYSTEM

KEY TERMS

peristalsis	exocrine	coenzyme
lacteal	bolus	duct
sphincter		

Imagine your favorite food. You may see it as a burger smothered in onions, a warm bowl of vegetable soup, a platter of pasta, or a slice of cherry cheesecake. Your body "sees" it as carbohydrates, lipids, and proteins. The body derives all the energy, structural materials, and regulating chemicals that it needs from the food that we eat.

CHEMICAL COMPOSITION OF DIET

Carbohydrates

When you hear the word carbohydrate, you probably think bread, potatoes and pasta. That's true, but there's more to it than that. Carbohydrates are composed of carbon, hydrogen, and oxygen. They include starches, glycogen, cellulose, and sugars. In humans, the principal role of carbohydrates is to provide a readily available source of chemical energy to generate an energy source (ATP) that drives metabolic reactions. Some sugars provide the building blocks of the nucleic acids. The main storage carbohydrate in humans is glycogen, which is stored in the liver and the skeletal muscles. Carbohydrates make up between two to three percent of your total body mass.

Cellulose, the most plentiful organic substance on earth, is ingested by humans, but is not digested. It creates fiber in our diet and helps move food and wastes through the gastrointestinal tract.

Lipids

Lipids include triglycerides (fats and oils), phospholipids, steroids, fatty acids, and some vitamins, such as vitamin E and vitamin K. Most are insoluble in polar solvents, like water, so they qualify for the term hydrophobic. The most plentiful lipids in your body are the triglycerides. Chemically, triglycerides are composed of glycerol and three fatty acids. They protect, insulate, and store energy in the body. Triglycerides provide more than twice as much energy per gram as either carbohydrates or proteins. They are stored in adipose (fat) tissue, and any excess of these food sources (carbohydrates, proteins, fats and oils) share the same fate.

Phospholipids are the major component of cell membranes. In addition to glycerol and two fatty acid chains, they contain a phosphate group. Steroids, composed of four rings of carbon, have a variety of functions: Cholesterol, for example, is a minor component of animal cell membranes and is the precursor of bile salts, vitamin D, and the steroid hormones, including the sex hormones and adrenocortical hormones (such as cortisol and aldosterone). Other lipid compounds of importance to humans include fatty acids, carotenes, vitamins E and K, and lipoproteins. Lipids make up eighteen to twenty-five percent of body mass in lean adults.

Proteins

As letters are to words, amino acids are to proteins. They are organic compounds composed of carbon, hydrogen, oxygen, and nitrogen; some contain sulfur. Proteins have a myriad of roles in the human body. They are involved in structure, regulation, contraction, immunology, transport, and catalysis. Proteins make up twelve to eighteen percent of the average adult.

Vitamins

Vitamins are organic nutrients required in small amounts to maintain growth and normal metabolism. Water-soluble vitamins (B vitamins and C) are absorbed with water in the digestive tract. Vitamins classified as fat-soluble (A, D, E and K) are absorbed in the digestive tract in the same manner as lipids. Most vitamins have roles as coenzymes, i.e., they are nonprotein molecules necessary for proper enzyme function.

- **Most vitamins cannot be made by the body and need to be ingested. Vitamin K is an exception; bacteria in the colon manufacture this vitamin important in the clotting of blood.**

VITAMINS

Vitamin	An Area of Importance	A Major Source
A	vision, skin, hair	dairy, green vegetables
D	bones and teeth	dairy products, tuna
E	red blood cell membrane	leafy vegetables, whole grains
K	blood clotting	cauliflower, cabbage
B1	carbohydrate metabolism	whole grains
B2	energy metabolism	milk, whole grains
B3	energy metabolism	whole grains, organ meats
B6	amino acid metabolism	whole grains, meat, fish
B12	red blood cell formation	dairy, meat
C	collagen formation	tomato, citrus fruit
Biotin	carbohydrate metabolism	eggs
Folic acid	formation of red blood cells, DNA	nuts, orange juice

Minerals

Minerals are inorganic elements that occur naturally in the earth's crust. They constitute about four percent of the human body and are concentrated most heavily in the skeleton. Minerals may be found in combination with each other, with organic compounds, or as ions in solution. The minerals present in the largest amount in the body are calcium, phosphorus, potassium, sulfur, sodium, sulfur, sodium, chloride, magnesium, iron, and iodide.

Water

You've probably heard someone say that the human body is mostly water. Actually, water constitutes up to sixty percent of adult body

MINERALS

Mineral	An Area of Importance	A Major Source
Calcium (Ca)	bones, teeth, muscle contraction	dairy, green vegetables
Chlorine (Cl)	water balance	table salt
Copper (Cu)	hemoglobin synthesis	legumes, seafood
Fluorine (F)	bones and teeth	tea, fluoridated water
Iodine (I)	thyroid hormone synthesis	table salt, seafood
Iron (Fe)	hemoglobin synthesis	legumes, eggs, whole grains
Magnesium (Mg)	protein synthesis	whole grains, green vegetables
Phosphorus (P)	bones and teeth	dairy, meat, green vegetables
Potassium (K)	muscle contraction	fruits and vegetables
Sodium (Na)	pH balance, nerve conduction	table salt
Zinc (Zn)	tissue growth, wound healing	legumes, meat, whole grains

weight. The figure is higher in children. Water needs vary depending upon diet, activity, environmental temperature and humidity. This makes it difficult to establish general water requirements. The obvious dietary source of water is by drinking it by itself, or as part of other beverages, but nearly all foods contain water. Thirst is the conscious desire to drink.

ORGANS AND ACCESSORY ORGANS OF THE DIGESTIVE SYSTEM

In humans, the gastrointestinal tract (alimentary tract) is a continuous nine-meter tube from the mouth to the anus. The organs of the digestive tract include the mouth, esophagus, stomach, small intestine, and large intestine. Accessory digestive organs include teeth, tongue, and salivary glands in the mouth. The pancreas, liver, and gallbladder are accessory organs to the small intestine (See Figure 13.1).

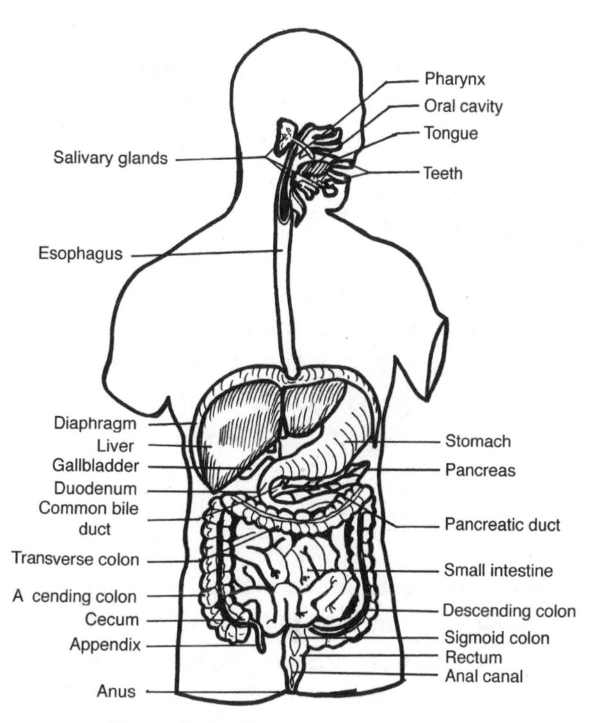

Figure 13.1—Human Digestive System

The digestive system prepares food for the cells through five basic activities: ingestion, movement, digestion, absorption, and defecation. Food enters the mouth by the process of ingestion. Once swallowed, food travels down the digestive tract due to peristalsis (contraction of smooth muscles lining the alimentary tract). Chemical digestion breaks down the food macromolecules to various monomers, which allows these substances to be absorbed from the small intestine into the cardiovascular and lymphatic systems. Indigestible substances are then eliminated from the large intestine in the process of defecation.

Mouth, Teeth, Lips, Tongue, and Salivary Glands

Digestion begins in the mouth, where food is mechanically broken down by the teeth. The teeth cut, tear, and pulverize food reducing solid foods to smaller particles for swallowing. The tongue is a muscular organ that maneuvers the food for chewing, shapes the food into a bolus (a chewed ball of food) and positions it for swallowing. Your taste buds, receptors for food stimuli, are positioned on the surface of the tongue. Saliva mixes with the ingested food to form the bolus. The production of saliva is under nervous control. Buccal cells (tiny glands in mucous membranes) produce some, but the three pairs of salivary glands (parotid, submaxillary, and sublingual) produce the majority of saliva.

- **Saliva is mostly water and, therefore, a good medium for dissolving foods so they can be tasted and passed on through the system.**

The solutes of saliva include bicarbonates and phosphates, salts, mucin and lysozyme. Bicarbonates and phosphates act as buffering chemicals for the food that enter the mouth. The salts activate an enzyme so starch digestion can begin. Carbohydrates, such as starch, are partially broken down in your mouth by the enzyme salivary amylase. Mucin is a protein of saliva that lubricates the food making it easier to swallow. Lysozyme is an enzyme that can destroy some bacteria, thereby serving as a defense against some microbes.

Esophagus

The only choice you have when it comes to digestion is what you eat. Once it is swallowed and the bolus reaches the back of the throat (pharynx), the process becomes involuntary. The bolus passes through a muscular, mucus-lined, collapsible tube called the esophagus. Peristalsis carries the food down the esophagus and into the stomach. A sphincter contracts at the end of the esophagus as food enters the stomach.

Stomach

The stomach is a perfect example of structure and function. It is a large sac-like, J-shaped organ with elastic walls with rugae (folds in the wall). This muscular organ enhances mechanical digestion by churning the food, while the mucus lining protects the organ from digesting itself.

- **An expanded stomach can hold approximately two liters of food.**

After the food has been in the stomach for several minutes, gentle, rippling, peristaltic movements, or mixing waves, pass every fifteen to twenty-five seconds. These muscular movements macerate the food, mix it with gastric secretions, and reduce it to a thin brownish liquid called chyme. Nerve impulses and the hormone gastrin control stomach secretions. Epithelial cells of the stomach lining are responsible for the various secretions. A thin mucus (mucin) protects the stomach lining from the very acidic conditions. The stomach is acidic due to the production of hydrochloric acid, which has three main functions:

- kill certain bacteria
- denature protein
- activate pepsin

The enzyme pepsin begins the hydrolysis (breakdown) of proteins. In response to stomach distension and the stomach hormone gastrin, nerve impulses regulate the emptying of the stomach. Food in the form of chyme is emptied into the duodenum (upper portion of the small intestine) between two and six hours after ingestion. The strongest mixing of the stomach contents occurs near the pyloric sphincter. This muscular ring regulates the passage of food from the stomach to the small intestine. Over the course of the digestion process, small amounts of chyme pass into the small intestine while most of the contents are pushed back into the stomach for additional mixing. Most of the substances are absorbed in the small intestine, but there is some absorption in the stomach of water, electrolytes, aspirin, and alcohol.

- **Carbohydrates spend the least amount of time in the stomach, followed by proteins and fats.**

Small Intestine, Pancreas, Liver, and Gallbladder

Approximately six meters in length, the small intestine is composed of three main segments:

- **The duodenum:** acid chyme from the stomach mixes with digestive juices from the pancreas, liver, gallbladder, and gland cells of the intestinal wall.
- **The jejunum:** primarily involved in the absorption of nutrients and water.
- **The ileum:** involved in absorption of nutrients and water. It joins the large intestine at the ileocecal valve, which controls passage of matter from the small intestine to the large intestine.

Segmentation in the small intestine allows for localized contractions in areas containing food, and brings food particles into contact with the lining of the small intestine for about fifteen minutes per segment. Peristalsis propels the chyme through the small intestine. Muscle contractions are much weaker here than in the esophagus and stomach. Chyme is in the small intestine for between three to five hours and it moves at the rate of one centimeter per minute.

The small intestine is site of most of the enzymatic hydrolysis of food and absorption of nutrients. The surface area of the small intestine is enhanced by its length and folding. The villi are projections from the inner wall of the small intestine. There are ten to twenty villi per square mm that increase the surface area of the intestine. Each villus has an arteriole, venule, and capillary network to facilitate nutrients entering the cardiovascular system. A single lacteal in each villus allows fats to enter the lymphatic system (See Figure 13.2).

Intestinal juice is a clear, yellow fluid with a slightly basic pH of 7.6. Two to three liters of the fluid are produced each day. It is mostly water and is the vehicle for absorption of substances from chyme, as they come in contact with the villi.

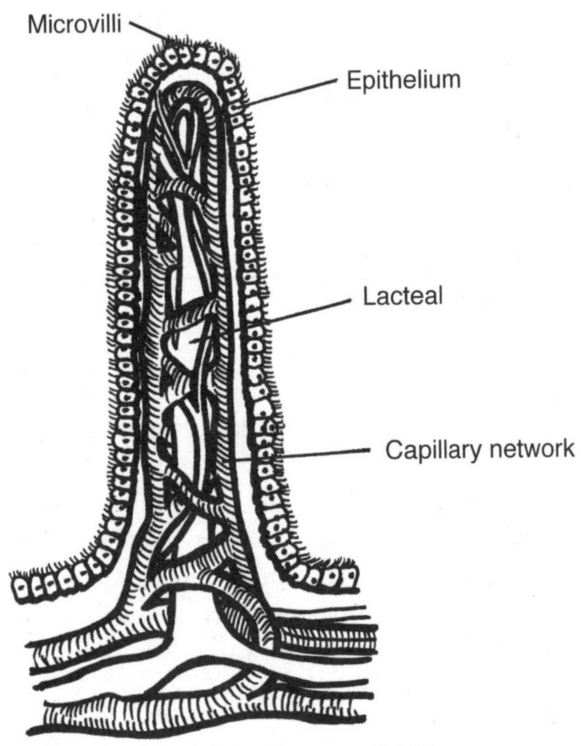

Figure 13.2—Structure of a Villus

The small intestine produces enzymes that break down disaccharides (carbohydrates) to monosaccharides: maltase, sucrase, and lactase. It produces peptidases to break down proteins to amino acids, and enzymes to break down nucleic acids: deoxyribonuclease and ribonucleases.

The pancreas is an accessory organ to the digestive system. It is connected to the small intestine by two ducts that enter the duodenum. The exocrine (ducted) portion of the pancreas (acini cells) produces a mixture of digestive enzymes called the pancreatic juice. This is a clear, colorless liquid containing water, some salts, sodium bicarbonate, and enzymes. The pancreas produces up to one and a half liters of this fluid daily. It stops the action of pepsin.

Before lipases can work, bile must emulsify the fats. The liver, the largest internal organ, produces bile and sends it to the duodenum via the common hepatic duct. Bile is a yellow, brownish, or olive green liquid composed of water and bile salts, cholesterol, lecithin, bile pigments, including bilirubin, and several ions. The bile salts break down fat globules into a suspension of tiny fat droplets. Their action is much like a detergent on greasy utensils. Up to one liter of bile is produced daily by the liver.

> The pancreas is a major player in the production of digestive enzymes. Pancreatic amylase continues the digestion of starch. Three enzymes involved in protein digest are trypsin, chymotrypsin, and pepsin. Deoxyribonuclease and ribonuclease are also manufactured by the pancreas and are involved in the breakdown of nucleic acids. Pancreatic lipase is involved in the breakdown of fats.

The gallbladder stores and concentrates the bile that is not used in the small intestine. The lipases of the intestine and the pancreas can now convert the fats to free fatty acids and monoglycerides. Once inside of the villi, they are resynthesized to triglycerides, coated with a protein cover and now called chylomicrons. These enter the lacteals (tiny lymph vessels) of the villi and drain into lymphatic vessels before draining into a vein in the neck region.

The catabolism of the food ingested is completed in the small intestine, and amino acids, monosaccharides, and nucleotides are absorbed into the circulatory system by way of the capillaries of the villi. The blood carries these digestive products via the hepatic portal vein to the liver (See Figure 13.3). The liver is in a position to chemically modify the products of digestion before they reach the rest of the body. The liver plays a key role in regulating the amount of glucose in the blood.

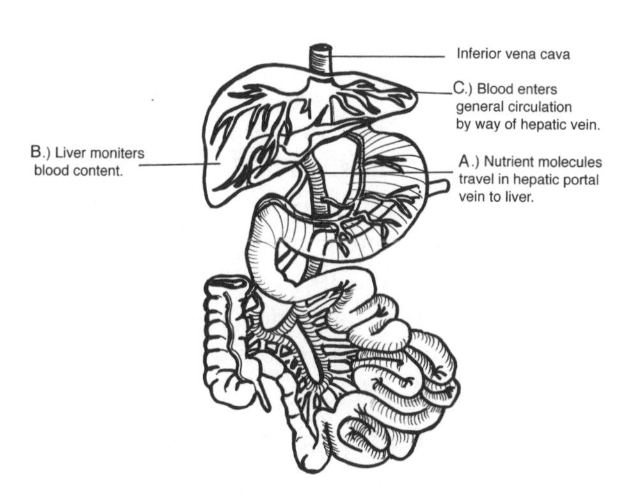

Figure 13.3—Hepatic Portal System

Large Intenstine

The small intestine empties directly into the large intestine or colon, and has the appearance of an upside-down U. There are no villi in the lining of the large intestine to increase surface area. Most of the fluid secreted into the alimentary tract is absorbed by the small intestine. The final absorption of water occurs in the large intestine. The peristaltic movement of undigested plant material (fiber) is sluggish. Material can remain in the large intestine for a period of up to twenty-four hours.

Chemical digestion in the colon is due to bacterial activity, not enzymes. Many bacteria, including *Escherichia coli,* live and reproduce in the colon. Some of the bacteria produce vitamins including biotin, vitamin K, folic acid, and several B vitamins. They ferment any remaining carbohydrates, and release hydrogen, carbon dioxide, and methane gas.

The wastes of the digestive tract become more solid as they pass through the large intestine. Feces contain large amounts of bacteria, as well as cellulose and other undigested material. Compacted feces pass from the large intestine into a short tube called the rectum. From the rectum, the feces leave the body through the anus. Two sphincters control the exit of feces. The first is composed of smooth muscle and is under involuntary control. The second is composed of striated muscle and is under voluntary control. This permits a conscious delay in defecation and allows storage of feces in the rectum.

DIGESTIVE SYSTEM DISORDERS

Ulcers

The human stomach has a hostile environment with its hydrochloric acid and low pH. The structure of the stomach is such that the organ and its underlying tissues are protected from these extreme conditions. A mucus lining is its first defense. In addition, the epithelial cells lining the stomach have tight junctions, which prevent leakage of the acid to tissues underneath.

- **The entire stomach lining is replaced every three days.**

If the lining is compromised, the gastric wall is injured by the stomach contents. This causes erosion of the stomach wall and peptic ulcers result. If gastric juice contents back up into the esophagus, erosion occurs in this area also. When the acid and pepsin break through the stomach wall, the acid stimulates the secretion of another protein, histamine. This further increases the production of hydrochloric acid and the cycle continues.

Until the 1990s eating a bland diet and taking antacids and antihistamines were the accepted methods of treating stomach ulcers. Recently, a bacterium, *Helicobacter pylori* was discovered to be the cause of about eighty percent of human ulcers. This motile bacterium with four to six flagella can burrow itself into the lining of the stomach and live in the mucosal wall, where it is protected from the highly acidic stomach environment. *H. pylori* secretes toxins which cause a persistent inflammation at the site of their colonization. This inflammation weakens the stomach wall. Antibiotics, which can completely cure bacterially-caused ulcers, are now the treatment of choice.

Colorectal Cancer

Colorectal cancer is among the deadliest of malignancies. Some individuals have a predisposition to this cancer due to their genetics. Increased risk of colorectal cancer has been associated with diets rich in animal fat, protein, and alcohol. Signs and symptoms of the disease include diarrhea, constipation, cramping, abdominal pain, and rectal bleeding. Tumors may have to be removed surgically.

Hepatitis

Hepatitis is an inflammation of the liver that can be caused by viruses, drugs, and chemicals, including alcohol. Depending on the virus, hepatitis can be mild or severe. Hepatitis A is an infectious form caused by a virus and is spread through fecal contamination. This is generally mild in both children and adults and most recover with no permanent liver damage within four to six weeks. On the other hand, the viral disease hepatitis D results in severe liver damage and is often fatal.

SUMMARY

- The oral cavity is the entrance to the digestive tract and the site of the ingestion of nutrients. Digestion in the mouth is minimal but the breakdown of carbohydrates begins here with the action of salivary amylase. Once the food is swallowed, peristaltic movements along the esophagus move the food into the stomach.
- The stomach is a muscular organ that churns and mixes the food into a liquid called chyme. Small amounts of the chyme enter the small intestine. The small intestine itself produces enzymes to break down all types of macromolecules.
- The amino acids and monosaccharides enter the circulatory system through the capillaries of the villi and the fats enter the lacteal of the villi and enter the lymphatic system. Most of the absorbed nutrients immediately pass through the liver for processing.
- The large intestine serves primarily to concentrate and store undigested food residues until they can be eliminated from the body as feces. The only chemical digestion going on in the large intestine is by bacteria, which also contribute to the production of some vitamins, including vitamin K and some of the B's.
- Most ulcers are caused by bacterial infections and can be cured with the proper antibiotic. Colorectal cancer is a deadly disease. Some people have a genetic disposition for this disease. Hepatitis is an inflammation of the liver. It has several causes including viruses, drugs, and chemicals, including alcohol.

THE CIRCULATORY SYSTEM

KEY TERMS

mast cell	endocytosis	exocytosis
pericardium	mediastinum	

Galen was the surgeon for emperor Marcus Aurelius of Rome around 200 AD. He believed that "natural spirits" in the liver transformed food into blood, which flowed to the heart, went to arteries, and then to the tissues where it was absorbed. Old blood, he speculated, was replaced by new blood formed in the liver. Veins were considered separate from the arteries. It was thought that blood just swished back and forth in the veins. This theory survived for 1400 years. You could say science was stagnant!

WILLIAM HARVEY

In 1628, William Harvey hypothesized that blood circulated through the body. Blood left the heart through the arteries, passed through tissues and returned to the heart through the veins. He demonstrated the one-way flow of the blood with a very simple experiment that is often repeated in elementary school classrooms today. He pressed his finger against one of the major veins in his forearm and moved his depressed finger along the vein towards his hand, pushing the blood out of the vein. If the vein only carried blood in one direction, i.e. back to the heart, the vein should remain empty. In fact, the vein remained empty until his finger was removed.

Some of the greatest science experiments are simple, yet convincing and elegant. The microscope needed to be invented before the connections between arteries and veins could be visualized. In 1661, Marcello Malpighi discovered capillaries by microscopically viewing lung tissue from frogs.

The circulatory system is composed of three components: blood, blood vessels, and the heart. Let's review them one by one.

BLOOD

Blood is the only liquid connective tissue in the body. The human adult has about 4.7 liters of blood. It has roles in transport, regulation, and protection. The blood transports oxygen, carbon dioxide, nutrients, hormones, heat, and wastes. It is involved in the regulation of body temperature, pH, and the water content of cells.

- **The body is protected from blood loss through clotting, and against disease by phagocytic white blood cells and antibodies.**

Plasma—Composition and Function

Plasma makes up fifty-five percent of the blood. It is a clear, straw-colored liquid which is mostly water, the blood's solvent. The plasma transports nutrients, waste products of metabolism, respiratory gases, and hormones. There are three main plasma proteins.

Albumin is the smallest and most numerous of the plasma proteins. It helps recover water that has been lost through the capillaries and transports some of the steroid hormones.

Immunoglobulins (antibodies) help in the immune system by attacking bacteria and viruses. Other globulins help in the transport of iron, lipids, and fat-soluble vitamins.

Fibrinogen is the third plasma protein and it has an essential role in the clotting of blood, providing the necessary protein network. Various ions are solutes in plasma. They play key roles in osmotic balance, pH buffering, and the regulation of membrane permeability.

Blood Cells—Structure and Function

A little less than half of the blood is comprised of cells. These include red blood cells used in the transport of oxygen to all cells, white blood cells involved in defense and immunity, and platelets which seal leaks in blood vessels.

CELLS OF THE BLOOD

Cell type	Function
erythrocytes (red cells)	transport of oxygen and carbon dioxide
leukocytes (white cells) neutrophil macrophage eosinophil basophil lymphocyte	defense and immunity
platelets	blood clotting

Red Blood Cells (Erythrocytes)

Why is shed blood red? Red blood cells or erythrocytes contain the oxygen-carrying protein hemoglobin, which is the pigment that gives blood its characteristic red color. They are the most numerous cells in your body (5-billion/ml blood) and the simplest. Mature red blood cells are flattened and disc-shaped with a central depression. Red blood cells could be described as non-reproducing sacks of oxygen-binding hemoglobin. Approximately 2.5 million erythrocytes are manufactured every second in the red bone marrow. The hormone, erythropoietin, triggers transformation of stem cells in the marrow to produce red blood cells. After circulating for about three to four months in the blood, red blood cells are engulfed by liver and spleen scavenger cells.

- **Mature red blood cells lack a nucleus, ribosomes, and mitochondria.**

White Blood Cells (Leukocytes)

Unlike red blood cells, the leukocytes contain a nucleus and do not have hemoglobin. Most leukocytes live only a few days, although some, especially lymphocytes, can live for several months or longer. During an infection, the white blood cells may only live for a few hours. The shape of their nuclei and the staining properties of their granules can distinguish white blood cells from one another. Five classes of white blood cells are recognized.

Neutrophils and *macrophages* are active in phagocytosis, ingesting bacteria and cellular debris. Certain chemicals released by bacteria and inflamed tissue attract the white blood cells to the site. After engulfing bacteria, neutrophils may release lysozyme, which destroys certain bacteria. They also release strong oxidants, such as peroxide, and proteins called defensins that have antibiotic activity. *Monocytes* arrive after the neutrophils and enlarge to become macrophages, which clean up cellular debris and bacteria after an infection.

Eosinophils enter tissue fluid from the capillaries and release enzymes to combat allergic reactions. *Basophils*, on the other hand, intensify the inflammatory response when they leave the capillaries and enter tissues. The basophils develop into mast cells, which liberate proteins involved in allergic reactions.

The *lymphocytes* are the major combatants in immune responses. They are the B cells, T cells, and natural killer cells. These cells are particularly active in fighting infections caused by viruses, bacteria, and fungi. They are also responsible for transfusion reactions, allergies, and the rejection of transplanted organs.

> The number and type of white blood cells can indicate a person's health. Most infections stimulate an increase in circulating white blood cells. Monocytes increase in individuals with mononucleosis. Infection with hookworm causes an increase in the number of eosinophils. HIV infection depletes certain white blood cells.

Platelets

Platelets are small, cell-like fragments derived from special white blood cells called megakaryocytes. They have no nucleus and live for approximately between five and nine days. Aged and dead platelets are removed by macrophages in the liver and the spleen. Platelets release chemicals involved in blood clotting.

Mechanism of Blood Clotting

Normally, blood stays in its liquid form as long as it stays within its vessels. A blood clot is a gel that contains formed elements of the blood entangled in fibrin threads.

Clotting is a complex cascade of events in which one clotting factor activates the next one in a fixed sequence (See Figure 14.1).

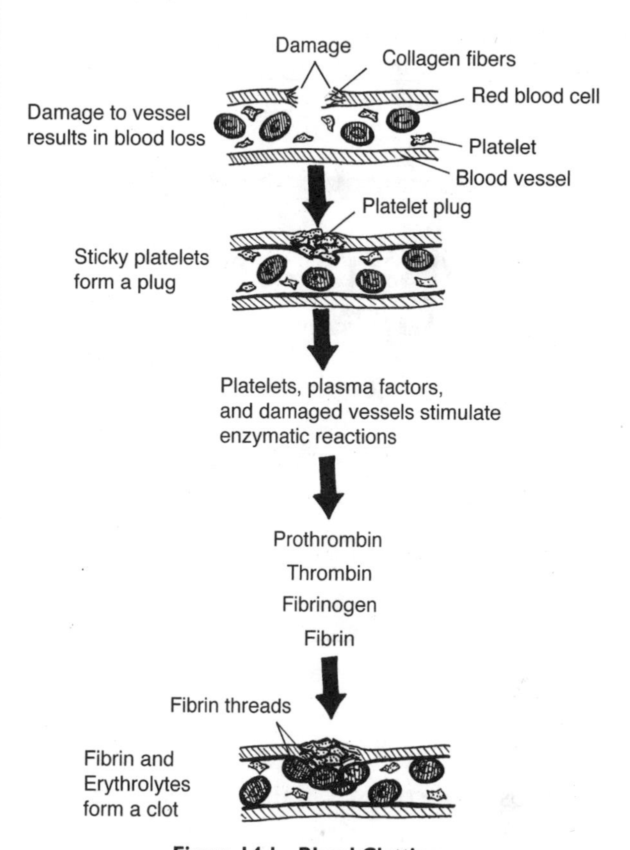

Figure 14.1—Blood Clotting

When a blood vessel is injured, platelets immediately become trapped at the injury site. This triggers constriction of the blood vessel by muscular contraction. Platelets release chemicals that make nearby platelets sticky, causing more platelets to adhere to each other and to collagen. A platelet plug is formed. Clotting factors secreted from platelets, damaged cells and factors in the plasma including calcium and vitamin K, begin the conversion to a fibrin clot. These factors are the catalysts for the conversion of prothrombin, a glycoprotein, to thrombin, an enzyme that is the catalyst for the conversion of fibrinogen to fibrin, which is the mesh-like clot. Blood clots prevent you from bleeding to death. The down side is that most heart attacks are due to a blood clot in a coronary artery already narrowed by plaque. Death of heart muscle is rapid and irreversible. Similar clotting in brain vessels is a cause of strokes.

BLOOD VESSELS—STRUCTURE AND FUNCTION

There are tens of hundreds of kilometers of blood vessels in the body. This ensures that every cell is within diffusion distance of a capillary. Five basic types of vessels that differ in form and function make up the circulatory system.

Arteries

These vessels have a large diameter and are composed of complex walls. The outer coat is loose connective tissue, the middle coat is a very thick layer of elastin and smooth muscle, and the inner lining is a layer of flattened interlocking endothelial cells (See Figure 14.2). As blood leaves the heart, it enters the main arteries. The walls of these vessels are pushed outward, increasing fluid capacity. As the walls stretch, the elastic fibers recoil and exert pressure on the blood. This is measured as blood pressure. As the blood is pushed out of the arteries, the diameters of the arteries decrease and the arterial pressure drops, until the next contraction abruptly returns the pressure to its maximum.

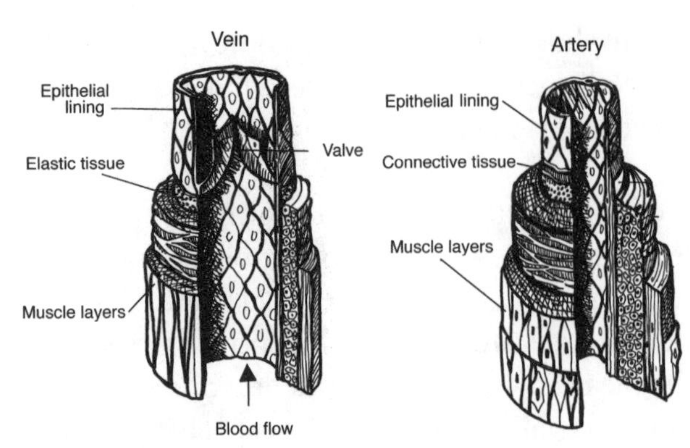

Figure 14.2—Structure of Arteries and Veins

- **Systolic pressure is the highest pressure attained in the arteries, and diastolic pressure is the lowest blood pressure before the next contraction.**

Arterioles

The major arteries branch into smaller and smaller arteries, eventually giving rise to arterioles. These are small vessels whose walls lack elastic fibers but contain a majority of smooth cells. The amount of blood that flows into a particular tissue depends largely on the diameter of the local arterioles. If an organ needs oxygen, the muscle cells of the arterioles relax and the arterioles increase in size, thereby increasing the blood to that organ. The reverse occurs to an organ during low activity; the arteriole muscles are contracted, which reduces the diameter of the vessel.

Capillaries

The capillaries are the smallest, shortest, and most porous vessels of the circulatory system. The structure of these vessels creates greater surface area for the exchange of materials between blood and the interstitial fluid. These microscopic vessels are a single layer of flattened cells whose edges fit together like pieces of a puzzle. Between these cells are the pores of the capillaries. Red blood cells and most proteins are too large to penetrate the pores.

- **There are about 40,000 km of capillaries in the adult human.**

What happens at the capillaries? Oxygen, nutrients, carbon dioxide, and wastes go between interstitial fluid and the bloodstream by simple diffusion. Water is forced out at the beginning of the capillary in response to the steep pressure gradient between the blood and the surrounding fluids. As the blood moves through, pressure drops rapidly and less water is pushed out of the pores. Water is drawn back into the other end of the capillary by osmosis because there are more dissolved materials in the blood than in the surrounding fluid. Some substances move in and out of the capillary by endocytosis (in) and exocytosis (out). If you sustain an injury, swelling (edema) is due to excess fluid leaving damaged capillaries.

Venules

When several capillaries join, they form small veins called venules. The smallest venules are porous and are the site where many phagocytic white blood cells emigrate from the bloodstream into inflamed or infected tissue. As venules become larger, they converge to form veins.

Veins

All veins bring blood towards the heart. The lowest blood pressure is found in these vessels. Although veins are composed of essentially the same three coats as the arteries, the relative thickness of the layers is different. The outer coat is the thickest and is made up of collagen and elastic fibers. The lumen of the veins is larger than that of comparable arteries (See Figure 14.2). Since most veins run upward, or against gravity, what allows this blood to reach the heart? Skeletal muscle activity, as in walking, and pressure exerted by breathing, squeeze veins and force blood to the heart. In addition, veins are equipped with flaplike valves which project from the vein wall, allowing movement of blood in one direction. Our legs bear the most weight. If the walls of the veins become distended and the valves fail, a condition known as varicose veins ensues.

HEART—STRUCTURE AND FUNCTION

Make a fist. That's about the size of the human heart. The heart is a muscular pump and the major organ of the circulatory system (See Figure 14.3). The pericardium surrounds the heart and confines it to its space in the mediastinum, while allowing sufficient freedom of movement for vigorous and rapid contraction. On average, the heart beats seventy-two beats per minute. Over a lifetime of seventy years, the heart beats approximately 2.5 billion times and pumps 200 million liters of blood!

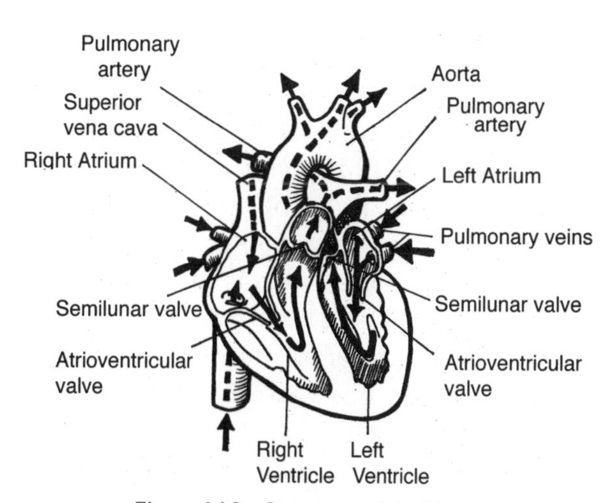

Figure 14.3—Structure of the Heart

- **The heart recirculates the entire volume of blood, almost five liters, every minute.**

The heart is a two-sided, four chambered, cone-shaped organ. On each side there is a thin-walled atrium which receives blood from veins and a thicker walled ventricle, which is larger and receives blood from the atrium above it. Blood leaves the ventricles and flows through arteries. The left and right sides of the heart are separated by a partition called the septum. Blood flows one way through the heart with the help of valves.

- **Left and right atrioventricular valves** are flaps of tissue that open in one direction. After the atria contract, pressure of the atrial blood increases and fluid pushes against the valves, which open, sending blood to the ventricles.
- **Semilunar valves** are flaps of tissue between the ventricles and the main arteries leaving the heart. They prevent a backflow of blood into the heart after the blood leaves the ventricles.

The sinoatrial node is specialized cardiac tissue in the right atrium and causes the heart to contract. From here a spontaneous electrical discharge originates every 0.6 seconds. A wave of electrical activity spreads across the atria, passing from one cardiac muscle cell to the next through communicating junctions (intercalated disks) that link neighboring cells. This causes a synchronous contraction of the atria. There is only one point of electrical connection between the atria and ventricles, and that is through the atrioventricular node. This arrives about 0.1 seconds later. This electrical wave spreads over the ventricles and causes them to contract in unison, forcing the blood into the major arteries. Areas of the hypothalamus and the medulla of the brainstem regulate the heart rate. The electrical activity of your heart can be read on an EKG. With each beat, the heart pumps blood into two closed circuits, the pulmonary circuit and the systemic circuit.

Pulmonary Circulation

The right side of the heart is the pump for the pulmonary circulation. It receives all of the deoxygenated blood returning from the systemic circulation. The blood enters the right atrium via the superior vena and inferior vena cava. Blood flows from the right atrium into the right ventricle and leaves the heart through the pulmonary trunk, which branches into the pulmonary arteries. Once the blood enters the capillaries of the right and left lungs, blood unloads carbon dioxide, which is exhaled, and picks up oxygen. The freshly oxygenated blood then flows into pulmonary veins and returns to the left atrium (See Figure 14.4).

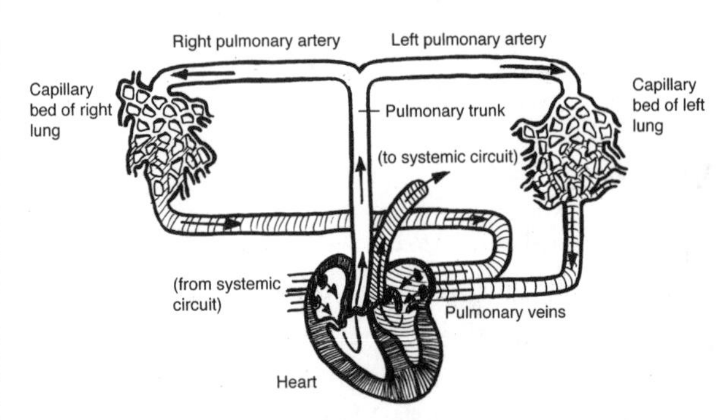

Figure 14.4—Pulmonary Circulation

Systemic Circulation

The left side of the heart is the pump for the systemic circulation. Newly oxygenated blood enters the left atrium, and then the left ventricle. The left ventricle ejects blood into the aorta. From this main artery the blood divides into separate streams, entering progressively smaller and smaller arteries that carry it throughout the body. In systemic tissues, arteries give rise to smaller diameter arterioles, which lead into extensive capillary beds. Here nutrients and gases are exchanged through the porous walls of the capillaries. In most cases, blood flows through only one capillary and then enters a systemic venule. These vessels carry deoxygenated blood away from the tissues and merge to form larger veins. The blood reaches the right atrium via the vena cava, a large vein (See Figure 14.5).

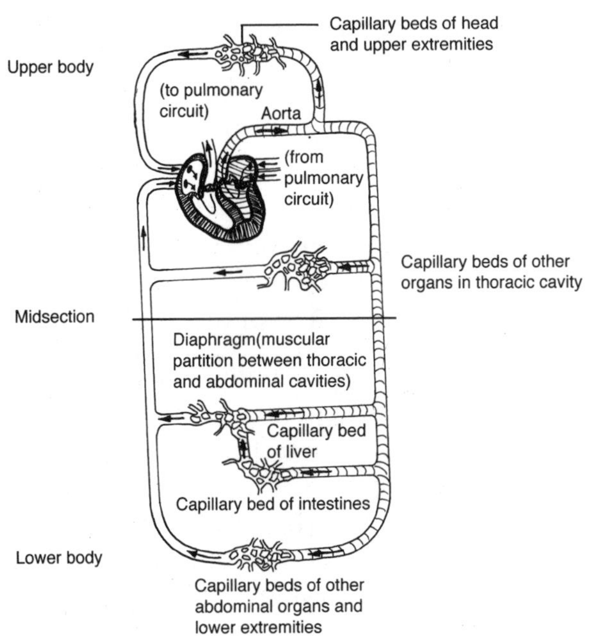

Figure 14.5—Systemic Circulation

CORONARY CIRCULATION

The right and the left coronary arteries branch from the ascending aorta and supply oxygenated blood to the cardiac muscle. Nutrients could not possibly diffuse from the chambers of the heart through all the layers that make up heart tissue. While the heart is contracting it does not receive much of the oxygenated blood. When the heart muscle relaxes, the high pressure of the blood in the aorta propels blood through the coronary arteries, into capillaries, and then into the coronary veins.

CIRCULATORY SYSTEM DISORDERS

Hypertension

Hypertension, or high blood pressure, is the most common disease affecting the heart and blood vessels. When pressure is high the heart uses more energy in pumping. Because of the increased effort, the heart muscle thickens and the heart becomes enlarged. It needs more oxygen. Untreated high blood pressure can lead to decreased diameter of arteries and arterioles causing atherosclerosis, strokes, heart attacks, and kidney failure. There is no single cause of high blood pressure but there are certain conditions that place an individual at higher risk. Behavioral risk factors include obesity, smoking, fatty diets, and overuse of alcohol. Genetic and physiological risks, such as kidney disease, play a significant role in this condition.

Many kinds of anemia exist. All are characterized by insufficient red blood cells or low hemoglobin. These conditions lead to fatigue and intolerance to cold, both of which are related to lack of oxygen needed for energy and heat production, and to paleness, which is due to low amounts of hemoglobin. In some cases adding additional iron or protein to the diet overcomes the condition.

Atherosclerosis

Atherosclerosis is a process in which fatty substances, especially cholesterol and triglycerides, are deposited in the walls of medium-sized and large-sized arteries. Smooth muscle cells of the arteries accumulate more of the cholesterol, and a plaque can be formed in the blood vessel. This can block the artery and prevent blood flow. If this clot becomes dislodged it can travel to other areas of the body and block blood flow.

Heart Attack

Most heart problems result from faulty coronary circulation. A heart attack, myocardial infarction, is the death of an area of cardiac tissue because of an interrupted blood supply. This may result from a plaque in the arterial wall. The severity of the attack is based on how much of the heart tissue is damaged. The heart will lose some of its strength after an attack.

> The most common brain disorder is a stroke (cerebrovascular accident). Common causes of strokes include blood clots, atherosclerosis, or ruptured brain arteries. A clot-dissolving drug called tissue plasminogen activator is being used to open up blocked blood vessels in the brain and heart.

Leukemia

Leukemia is a malignant disease characterized by uncontrolled production of immature white blood cells—cells that lack the genetic ability to reach maturity. Anemia and bleeding are commonly seen due to the crowding out of normal bone marrow cells by the overproduction of immature cells, preventing normal production of red blood cell and platelets. The most frequent cause of death, however, is uncontrolled infection due to the lack of normal, mature white blood cells.

SUMMARY

- The circulatory system delivers oxygen and nutrients to the body's tissues and carries carbon dioxide and other wastes away.
- Blood contains cells and cell fragments. Red blood cells carry oxygen to the cells. White blood cells have roles in phagocytizing intruders and tissue debris, and in launching the immune response. Platelets aid in clotting of blood. Plasma is the liquid portion of the blood.
- The design of the blood vessels is well suited to their function. The large elastic arteries allow them to snap back after each surge from the heart, squeezing blood to the next destination. Muscular arterioles either increase or decrease in diameter, thus serving the needs of the tissues.
- The heart is a double pump. The right side of the heart pumps blood to the lungs for oxygenation. The left side of the heart receives the blood from the lungs and sends it via the aorta to all parts of the body. Valves in the heart keep the blood flowing in one direction.
- Although the heart will contract rhythmically without an external stimulus, neural impulses regulate the rate of contraction by stimulating the sinoatrial node, which initiates contraction of the atria and relays the stimulus to the remainder of the heart.

CHAPTER 15

IMMUNE SYSTEM

KEY TERMS

pathogen	specific	interferon
phagoctye	nonspecific	lymphocyte

You might not consider a trip to the grocery store, or dining at a local restaurant, a health threat, but there may be unseen dangers. As you go about your business each day, you come into contact with thousands of pathogens—disease-causing organisms including bacteria, fungi, viruses, and parasites.

Fortunately, your body is equipped with a variety of defenses against these invaders.

LYMPHATIC SYSTEM

You might be surprised to learn that your body has a second circulatory system—a system of vessels transporting a fluid called lymph throughout your body. Your lymphatic system (See Figure 15.1) filters lymph through small, lima-bean shaped structures called lymph nodes which contain *phagocytes* and *lymphocytes*, two types of cells that develop in bone marrow; they attack pathogens and are discussed later in this chapter.

The thymus gland, located just above your heart, is where specialized lymphocytes called *T cells* (see page 121) develop. This gland reduces in size after adolescence, once the immune system is mature.

The spleen, a fist-sized mass of tissue in the upper abdomen, has two sections—the white pulp, which stores lymphocytes, and the red pulp, which filters and stores blood.

DEFENSES AGAINST INFECTION

The body has three basic defense systems:

- nonspecific defenses
- nonspecific immune defenses
- specific immune defenses.

Nonspecific Defenses

Your body uses a variety of simple defenses to keep out many disease-causing invaders. Because these defenses do not target a specific pathogen, they are called *nonspecific defenses*. These include barriers such as the skin and mucous membranes, secretions such as tears, saliva, and mucus, and body processes including coughing, sneezing, and vomiting.

Nonspecific Immune Defenses

If invading pathogens are not deterred by the defenses above, the immune system takes action in several ways. *Nonspecific immune defenses*, or those that do not target a specific pathogen, include *phagocytes*, cells that engulf and consume invaders, and *interferons*, proteins secreted by infected cells that limit the harmful effects of viruses.

An easily recognizable defense is the inflammatory response, where redness, warmth, and swelling occur at an area of injury or infection. These symptoms mean that nearby blood vessels have dilated to increase blood flow to the affected area, increasing the action of phagocytes and speeding healing.

Specific Immune Defenses

Specific immune defenses are specialized responses that target a specific invader. The circulatory, lymphatic, and other systems coordinate in complex ways to target specific pathogens, but *lymphocytes* play the most critical role. These are specialized white blood cells found in high concentrations in the lymphatic system; when inactive they are stored in the white pulp of the liver.

If a lymphocyte detects an invader, it begins a series of cell changes, including cell replication and chemical communication with other immune system components.

Specific immune responses include:

- **mucosal or local immunity**—defense proteins called antibodies are produced in the tissues underlying the linings of the digestive and respiratory tracts to target and kill nearby pathogens.
- **cell-mediated immunity**—phagocytes engulf and partially digest pathogens, then specialized lymphocytes called T cells recognize and destroy both the phagocyte and the pathogen it contains.
- **antibody-mediated immunity**—lymphocytes produce antibodies that travel through the body via the circulatory and immune systems, targeting and neutralizing specific pathogens.

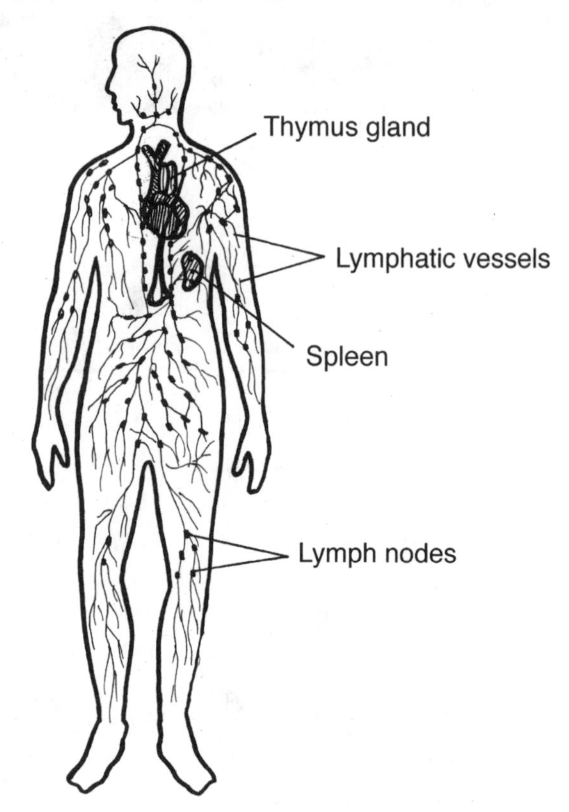

Figure 15.1—Human Lymphatic System

In addition to these responses, your body has another specific but adaptive immune response called the *anamnestic response* that helps it to recall invaders it has encountered in the past. When the body encounters a pathogen multiple times, this response triggers special memory cells that help to speed response time. This happens only when immunity has been actively acquired, either as a result of natural infection or vaccination.

Immune Cells at Work

As the chart at right shows, the cells of the immune system work together in a complex fashion to initiate specific immune defenses.

CELLS OF THE IMMUNE SYSTEM

Cell Type	Function
Helper T cell	starts the immune response, detects infection, and sounds the alarm, initiating both T cell and B cell responses
Inducer T cell	not involved in the immediate response to infection; mediates the maturation of other T cells in the thymus
Cytoxic T cell	detects and kills infected body cells; recruited by helper T cells
Suppressor T cell	slows the activity of T and B cells; scaling back the defense after the infection has been checked
B cell	precursor of plasma cell; specialized to recognize specific foreign antigens
Plasma cell	biochemical factory devoted to the production of antibodies directed against specific foreign antigens
Mast cell	imitator of the inflammatory response, which aids the arrival of leukocytes at a site of infection; secretes histamine and is important in allergic responses
Monocyte	precursor of macrophage
Macrophage	the body's first cellular line of defense; also serves as antigen-presenting cell to B and T cells and engulfs antibody covered cells
Natural killer cell	recognizes and kills infected body cells

Lymphocytes, the key players in immune defense, can be divided into two types:

- T cells, which mature in the thymus and are transported throughout the body, where they facilitate cell-mediated immunity by targeting and neutralizing pathogens;
- B cells, which facilitate antibody-mediated immunity by producing defensive proteins called antibodies, which circulate throughout the body to target and destroy pathogens.

In healthy individuals, both T cells and B cells have the ability to distinguish between your cells and those that don't belong to you. All cells contain *antigens*, molecules that allow lymphocytes to tell the difference. Cells that are yours contain *self-antigens*; foreign cells contain *nonself-antigens*.

As mentioned earlier, one of the body's first lines of defense involves phagocytes, cells that simply engulf and consume invaders. Among these are *macrophages*, cells critical in cell-mediated immunity. Macrophages engulf and only partially digest pathogens, but they retain the nonself-antigens that alert T cells to destroy the pathogen.

The immune system includes *cytotoxic T cells*, also called killer T cells, which are the first to recognize that other body cells are infected. Under routine conditions, cells of the body degrade internal cellular proteins on a regular basis; this might happen because the protein deteriorates, because the protein was incorrectly folded, or in the course of normal cell metabolism. Some of these are broken up into short peptides, which are then presented on the surface of normal body cells. If a virus or other pathogen is infecting a cell, then the pathogen, too, will produce proteins (called antigens), and eventually present these non-self antigens on the surface of cells.

Cytotoxic T cells recognize specific antigens. The response of T cells to these molecules is complicated and requires a second signal to the T cell, via co-stimulatory cytokines. There are several different sets of such co-stimulatory molecules. Among the challenges of modern science is to understand exactly how they interact, and to find new therapies that may control or stop immune system attack on self-tissues or organs.

- **Destroying the infection also destroys the host cell, sometimes causing further disease.**

Immune System Disorders

Allergies

The immune system usually distinguishes between pathogens and other harmless particles or chemicals in the environment. Sometimes, however, a person's body will react to a harmless substance as if it were an antigen, producing antibodies. The antibodies stimulate macrophages to release histamine. This causes blood vessels to expand, the eyes to produce tears, and the nasal passages to secrete mucus.

- **Antihistamines can often counter the effect of an allergic reaction.**

In a severe attack, the blood vessels expand so much that blood pressure falls dangerously low and breathing may become difficult. This is called anaphylactic shock and it can be life-threatening. An injection of epinephrine is a typical treatment.

Allergic reactions are also called hypersensitivity reactions. The mechanisms by which the immune system defends the body, and by which a hypersensitivity reaction injures it, are similar. An allergic reaction usually refers to reactions that involve antibodies of the immunoglobulin E (IgE) class. These antibodies bind to special cells, including basophils in the circulation and mast cells in tissues. When IgE antibodies are bound to those cells, and they encounter allergens, they are prompted to release chemicals that can injure surrounding tissues.

The best course is to avoid the source of an allergic reaction. When this isn't possible, allergen immunotherapy (injections) might provide an alternative. Measured amounts of the allergen are injected until a maintenance level is reached. This stimulates the body to produce blocking or neutralizing antibodies that may help to prevent a reaction.

Autoimmune Diseases

The immune system is a complicated network of cells and cell components that usually work to defend the body and eliminate infections caused by bacteria, viruses, and other invading microbes. A weakened immune system may result in chronic infections or even an immune system disorder. A disordered immune system targets the cells, tissues, and organs of a person's own body.

Lupus is an example of an autoimmune disorder that affects various organs and tissues. The brain is attacked in multiple sclerosis; in Crohn's disease, it is the gut that is attacked. Eventually, damage becomes permanent. For instance, destruction of insulin-producing cells in the pancreas in Type 1 diabetes mellitus is irreversible.

Although autoimmune diseases are rare, certain populations are affected more severely. Women appear to be at greater risk than men. Lupus is more common in African-American and Hispanic women than Caucasian women of European ancestry. Rheumatoid arthritis and scleroderma affect more Native Americans than the general population of the United States.

Genetic inheritance may predispose a person to autoimmune disease. Family members could inherit and share abnormal genes, but each develop a different type of autoimmune disease. Genes determine the variety of MHC molecules that an individual carries on the cells. They also influence the array of T cell receptors on T cells. Genes are not the only contributing factor, however, because some people with disease-associated MHC molecules do not develop an autoimmune disease.

Diagnosis can be difficult with chronic autoimmune disease, and the course of the disease can be unpredictable. Sometimes a patient can go for long periods without specific symptoms.

HIV/AIDS

Acquired Immune Deficiency Syndrome, (AIDS) is a condition in which the body is unable to protect itself against pathogens. AIDS is the result of human immunodeficiency virus, or HIV, which leaves the body unable to fight infection.

HIV replicates inside helper T cells, or T4 cells, finally destroying them. Since these cells stimulate the production and activity of B cells and killer T cells, their destruction means that HIV interferes with both humoral and cell-mediated immunity. The body ends up with no way to defend itself.

There is no definitive way to determine exactly how HIV/AIDS started, and when or where. There is some evidence of the disease prior to 1970. By 1980, HIV had spread to at least five continents. Today, HIV is the fastest-growing epidemic in the world.

HIV is spread by contact between people, specifically through the exchange of certain body fluids such as blood, semen, or breast milk. Infection may occur through sexual intercourse, through the sharing of needles among intravenous drug users, or by the transfusion of contaminated blood products. It can also cross the placenta from mother to fetus, and be transmitted to newborns through breastfeeding.

Healthy cells with CD4 receptors are usually called CD4-positive (CD4+) cells or helper T lymphocytes. These cells activate and coordinate other cells of the immune system such as B cells, macrophages and cytotoxic T cells, all of which help to destroy cancerous cells and invading organisms. The HIV virus attaches itself to lymphocyctes that have a receptor protein, called CD2, in the outer membrane. Via viral RNA, it is incorporated into the DNA of the infected cell where it reproduces itself, eventually destroying the cell and releasing new virus particles. These particles move on to other lymphocytes and the cycle continues.

- **People infected with HIV lose helper T cells in stages. Years may pass during which the HIV infected person has a slowly declining, below normal count of CD4+ cells.**

Some people develop flu-like symptoms, or symptoms like those of mononucleosis, a few weeks after contracting HIV. Further symptoms may not appear for years, although the virus circulates in the blood and other body fluids. Symptoms of AIDS include swollen lymph nodes, weight loss, a fever that comes and goes, fatigue, diarrhea, anemia and thrush (an oral fungal infection).

Treatments are particularly effective when drugs are given in combinations. Nucleoside reverse transcriptase inhibitors (such as AZT), non-nucleoside reverse transcriptase inhibitors, and protease inhibitors all help to slow the virus. HIV usually develops resistance to these drugs, but research continues on new drug therapies and regimens that have done much to extend and improve the quality of life for people with both HIV and AIDS.

PROGRESSION OF HIV

1. HIV virus attaches to and penetrates target cell.
2. HIV's RNA, the genetic code of the virus, is released into the cell. The RNA must be converted into DNA by the enzyme called reverse transcriptase.
3. Viral DNA enters the cell's nucleus.
4. Viral DNA integrates with the cell's DNA.
5. DNA now replicates and reproduces RNA and proteins in the form of a long chain that must be cut into pieces after the virus leaves the cell.
6. A new virus is assembled from RNA and pieces of protein.
7. The virus buds through the cell membrane, wrapping itself in a fragment of the envelope.
8. Viral enzyme (HIV protease) cuts structural proteins within the budded virus to be rearranged into the mature form of HIV.

SUMMARY

- The immune system defends the body against pathogens. There are three basic defense systems: non-specific defenses, non-specific immune defenses, and specific immune defenses.
- Specific immune defenses include mucosal or local immunity, cell-mediated immunity, and antibody-related immunity.
- Most immune system cells are white blood cells. There are many types, including B cells and T cells. The response of lymphocytes (or B and T cells) is complicated and requires signals within the body.
- Allergies result when the body reacts to an antigen by producing antibodies, even though the substance may be benign. Antihistamines can counter this effect. Sometimes allergen immunotherapy helps the body to keep reactions under control.
- Autoimmune diseases occur when a person's immune system attacks its own body cells, tissues, or organs. Autoimmune diseases are chronic, but often treatable, and can be managed successfully for long periods of time.

- HIV/AIDS inhibits the body from protecting itself against pathogens. The virus is spread by contact between people, specifically through the exchange of certain body fluids such as blood, semen, or breast milk.
- The virus enters cells, such as lymphocyctes, is incorporated into the DNA of an infected cell, and eventually reduces the number of CD4+, or helper T lymphocyctes. People with HIV often develop resistance to the drugs, so new combinations are always needed to battle this disease.

RESPIRATORY SYSTEM

KEY TERMS

perfusion	ethmoidal	mucous
bronchitis	allergen	

Hold your breath for a moment and you will become acutely aware of your respiratory system. As it performs its functions, the respiratory system is constantly interacting with the outside environment. It takes in air containing oxygen that cells and organs require, and expels the main end-product of metabolism, carbon dioxide. In the course of a single day, the lungs facilitate gas exchange between the 8,000 to 9,000 liters of air inspired and the10,000 liters of blood pumped through the pulmonary circulation by the heart!

STRUCTURE AND FUNCTION

Simply put, the lungs enable gas exchange between the air around us and the internal environment surrounding the cells in our body. The lungs expand when the diaphragm and muscles between the ribs (the intercostals) contract and enlarge the chest cavity. Similarly, the lungs deflate passively when the diaphragm and intercostals relax and reduce chest volume (See Figure 16.1). To facilitate these changes in lung volumes, the lungs are enclosed within a slippery membrane called the pleura, which lubricates and protects the surfaces of the lungs.

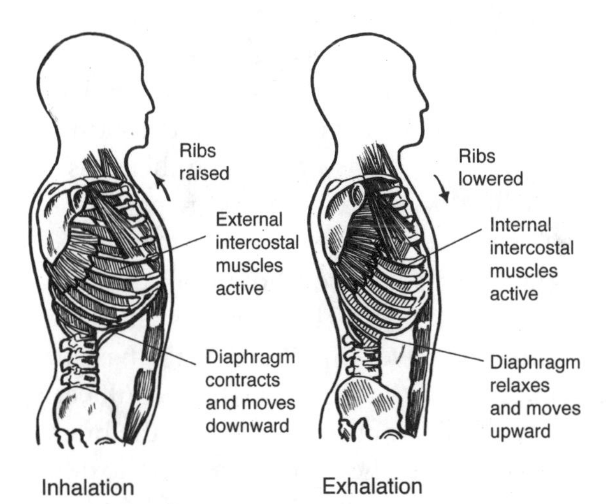

Figure 16.1—Breathing

The Airways

As the lungs begin to expand, air enters through the nose and mouth and passes into the respiratory airways. These airways are part of the upper respiratory tract that includes the nose, nasal cavity, ethmoidal air cells, frontal sinuses, maxillary sinus, larynx (voice box) and trachea. A flap of tissue called the epiglottis covers and protects the larynx as you swallow, keeping food particles from entering your respiratory airways. The trachea, or windpipe, branches into two smaller airways (bronchi) that ventilate the two lungs (See Figure 16.2). The bronchi themselves divide many times like branches on a tree. The smallest airways are bronchioles, which are only 500 μm in diameter.

Alveoli

At the end of each bronchiole are dozens of bubble-shaped, air-filled cavities called

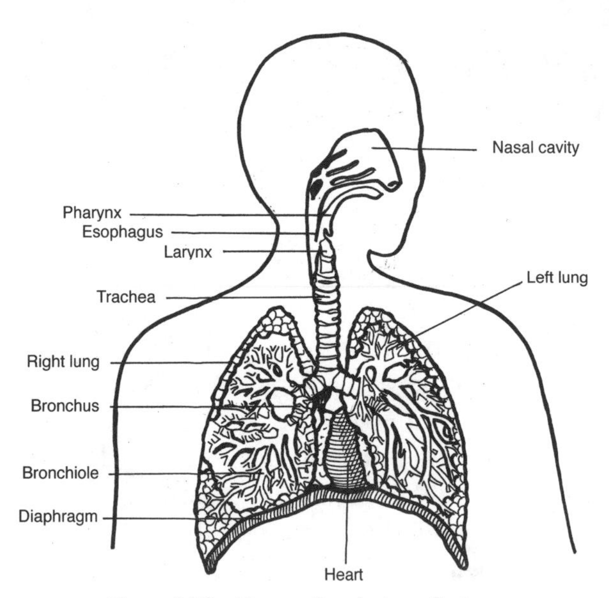

Figure 16.2—Human Respiratory System

alveoli that look like tiny bunches of grapes. The alveoli are really just little air sacs surrounded by capillaries (See Figure 16.3). This structure allows oxygen to move from the alveoli into the blood, and carbon dioxide to move from the blood into the alveoli. The amount of blood perfusing the capillaries of each alveolus is carefully matched to the amount of air flowing into the alveolus through the bronchioles. This matching of ventilation and perfusion assures that energy is not wasted by sending blood to an alveolus with a blocked airway, or by sending air to an alveolus with a compromised air supply.

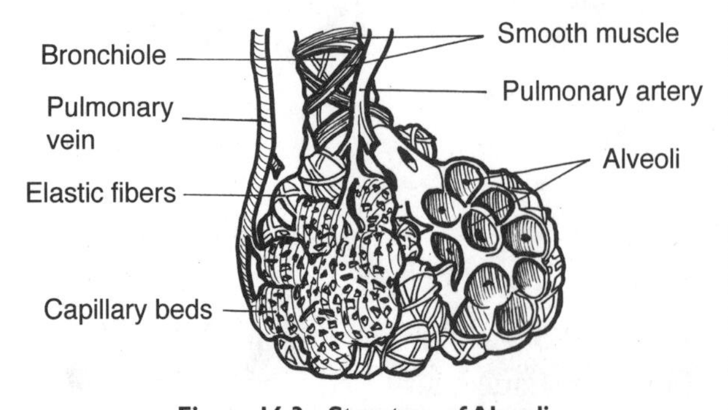

Figure 16.3—Structure of Alveoli

- **Although the lungs are internal organs, with each breath alien substances enter the body. These may include pollen, dust, smoke particles, viruses, bacteria, and whatever is carried by the air in the local environment.**

The maximum volume of air that can be inhaled and exhaled is called the *vital capacity* of the lungs and is up to five liters. Even when you exhale completely, you cannot eliminate all the air in the airways of the lungs (which would collapse them). This residual volume is called the *ventilatory dead space* and is usually about one liter in size. The amount of air that you breathe in and out with each breath is called the *tidal capacity*, which averages more than a liter in most individuals. On average, you usually take twelve to fifteen breaths per minute at rest, but your respiratory rate can increase to well over twenty breaths per minute during exercise.

REGULATION OF RESPIRATION

Breathing usually operates without us having to think about it, and is thus regarded as an "involuntary reflex." To assure that the rate of ventilation is matched to the oxygen demands of the body, breathing is closely regulated. Specialized nerve endings in the aorta and carotid arteries sense the chemical composition of the blood flowing through them. As you might expect, these sensors are particularly sensitive to changes in blood oxygen content, and also to the amount of carbon dioxide in the blood. These sensors send signals through nerves to a region at the base of the brain called the medulla. The medulla also receives signals from sensors within the brain itself that indicate how much carbon dioxide is in brain tissues. When levels of carbon dioxide start to rise, or when levels of oxygen start to fall, which happens whenever we increase our metabolism (during exercise, for example), the medulla sends signals to the respiratory muscles in the ribs and diaphragm. It instructs them to contract more strongly and more often, so that breathing and ventilation of the lungs increases. At the same time, the heart is instructed to pump blood faster through the lungs. These simultaneous events act to increase the amount of oxygen entering the blood and going to the tissues, and also increase the removal of carbon dioxide from the blood.

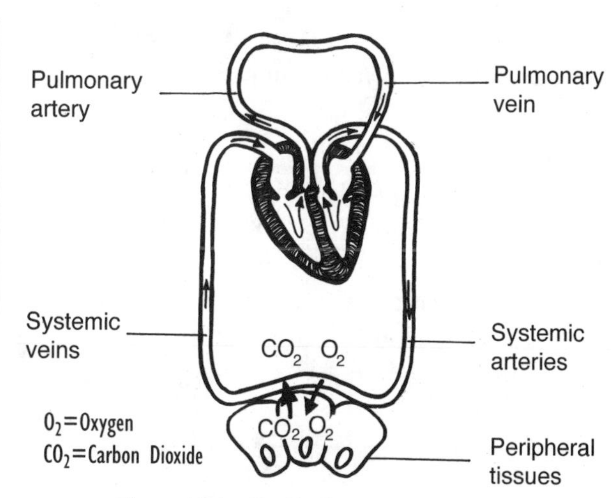

Figure 16.4—Gas Exchange in the Lungs

Hemoglobin

When oxygen enters the bloodstream from the alveoli, it quickly combines with a highly specialized molecule called hemoglobin. The main job of hemoglobin is to increase the ability of blood to carry oxygen, and it does this job so well that equal volumes of blood and air contain almost the same amounts of oxygen! Hemoglobin does this by tightly binding oxygen. However, when hemoglobin arrives in the tissues where oxygen levels are low and carbon dioxide levels are high, these conditions signal hemoglobin to dump its oxygen cargo. Once its oxygen is dumped, hemoglobin then picks up and helps carry carbon dioxide back to the lungs (See Figure 16.4).

- **The mighty molecule hemoglobin contains the element iron, which is a big reason we need iron in our diet. The iron in hemoglobin is also the reason our blood is red.**

Hemoglobin is carried inside red blood cells, also known as *erythrocytes*. As you might imagine, hemoglobin is very abundant in blood—100 ml of blood can contain as much as thirty-four grams of hemoglobin. Hemoglobin is also the reason carbon monoxide is so poisonous. Carbon monoxide binds hemoglobin much more tightly than oxygen does, which destroys hemoglobin's ability to carry oxygen.

The Big Picture

Overall, there are three main components of respiration: external, internal, and cellular. External respiration involves the gas exchange that occurs in the lungs. Internal respiration is the gas exchange that takes place between blood and body cells in tissues outside the lungs. Cellular respiration includes the processes inside cells that combine oxygen with many different fuels to obtain the energy necessary to sustain life.

Clearly, the respiratory system is essential for all three components of respiration. It achieves these functions with a high degree of specialization among the different tissues that make up the respiratory system. The airways, alveoli, and blood vessels that make up the lungs enable oxygen and carbon dioxide to move into and out of the body. Many other tissues and cells play protective roles. Some cells guard against invading pathogens, others scavenge foreign matter from the airways, some move tiny waves of mucous across the membranes lining the airways, and still others contain enzymes that act on substances carried by the blood that are crucial to blood-pressure control. With this specialization, the many jobs of the lungs are each performed at optimum efficiency so that gas exchange occurs reliably with minimum energy expense, and with minimum risk for infection.

RESPIRATORY DISORDERS

Asthma

Asthma is a chronic condition that affects more than 10 million Americans, and deaths due to asthma have increased dramatically in recent years. Asthma is caused by a sudden constriction of the respiratory airways. In very bad attacks, the airways can become inflamed, swell, and secrete large amounts of mucous. These attacks can be triggered by pollens, dust mites, animal dander, smoke, cold air and exercise, but do not appear due to emotional factors like anxiety or nervous stress. The factors that trigger attacks can be very different in different individuals.

The long-term symptoms of asthma also vary considerably among different individuals. Some have a chronic cough, other have a characteristic "wheeze," and others are troubled by recurrent bronchitis. Asthma also varies widely in frequency and severity. It is most common in children before the age of five, but also appears in middle-aged adults and seniors. In all these groups, it is most prevalent in residents of urban areas. Families with medical histories of asthma and allergies are also at high risk for asthma.

Asthma is most often diagnosed by specialized laboratory tests that evaluate airway function. Treatment for asthma usually involves reducing exposure to common allergens. Drug treatments are also available, including airway relaxants that provide immediate relief during an attack, as well as steroids that can "quiet down" airway sensitivity over the long term.

- A spirometer measures lung volumes and can help determine how well the lungs receive, hold, and move air. This tool can monitor lung disease, efficacy of treatment, and determine whether the lung disease is restrictive or obstructive.
- A peak flow meter measures the fastest speed at with which a person can empty the lungs, which can be important in evaluating how well the disease is being controlled.
- Chest X-rays can pinpoint where fluid has accumulated or where abnormal growths have occurred.
- Blood tests analyze the amount of carbon dioxide and oxygen in the blood to see how effectively the respiratory system is working.

Emphysema and Chronic Obstructive Pulmonary Disease

Emphysema occurs when the airway walls lose their structural support. Bronchioles in lungs with emphysema are like wet straws that collapse when you try to drink through them. When exhalation begins, the elevated pressure in the lungs collapses the affected airways, which traps air in the lungs and makes it very difficult to breathe.

- **Emphysema doesn't develop suddenly, but occurs gradually. Smoking, exposure to air pollution, irritating fumes and dusts on the job can cause the disease.**

Emphysema is diagnosed by pulmonary function tests, such as the spirometer and peak flow monitor, and also by blood tests, chest X-rays, sputum cultures, and electrocardiograms that record the electrical activity of the heart and shows abnormal rhythms that are typical in emphysema patients. Emphysema varies considerably in severity, can sometimes be treated with bronchodilator drugs, but is almost always irreversible and permanent.

Chronic obstructive pulmonary disease (COPD) is persistent obstruction of the airways caused by emphysema or chronic bronchitis. In this condition some alveoli may be destroyed, collapsed, stretched or over inflated. Such damage to the alveoli significantly decreases respiratory function and is usually permanent and irreversible. Early symptoms include shortness of breath and persistent cough, and can be associated with fatigue, anxiety, sleep problems, heart problems, weight loss and depression. Some of these symptoms can be treated to improve quality of life. Most common is inhalation oxygen therapy. Exercise can also improve quality of life in these patients.

World Trade Center Pulmonary Syndrome

World Trade Center Pulmonary Syndrome is a new condition, dating specifically from the attack on the World Trade Center on September 11, 2001. As the towers collapsed in New York City, everything within miles was coated with fine particles of dust that came from the buildings and the airplanes that struck them. This intense and immediate exposure to dust and smoke caused eye, nose, throat and lung irritation, and triggered coughing and sneezing, the body's usual way of removing foreign substances from the respiratory airways.

Workers and volunteers who returned to the site again and again experienced persistent symptoms including sore throat, hoarse voice, chest tightness, shortness of breath, persistent cough, wheezing, and continued shortness of breath. Other symptoms included nasal or sinus congestion, runny nose, facial pain, headache, nasal discharge, persistent throat clearing or cough, and postnasal drip. Gastrointestinal symptoms included an inability to tolerate certain foods, chronic indigestion, and a burning sensation in the chest. A host of psychological symptoms accompanied many of these respiratory and gastrointestinal symptoms.

Workers who had to return to the site during the recovery process were advised to wear protective clothing, eye goggles, gloves, and shoes and clothes designed to keep the dust away. Even with these precautions, many people developed a cough and others developed bronchitis or asthma. The causes for this pattern of symptoms remains unclear and remains under investigation.

According to some news reports, nearly half of the firefighters who worked at the World Trade Center site immediately following the disaster were coughing so severely that they required medical attention. Some suffered from rare lung inflammations called allergic alveolitis.

Lung Cancer

Lung cancer usually starts in the lining of the bronchi, but can also begin in other areas of the respiratory system, including the trachea, bronchioles or alveoli. It can also metastasize from other parts of the body. Cancers frequently spread to the lungs from other parts of the body by being carried through the bloodstream.

Most lung cancers are carcinomas, a cancer that begins in the lining or covering tissues of an organ. Tumor cells grow and spread and each type requires a unique treatment. Most lung cancers are called bronchogencic carcinoma, and include squamous cell carcinoma, small cell (oat cell) carcinoma, large cell carcinoma, and adenocarcinoma. Alveolar cell carcinoma originates in the alveoli. Less common lung tumors are bronchial adenoma, (cancerous or noncancerous) chondromatous hamartoma (noncancerous), and sarcoma (cancerous). Lymphoma is a cancer of the lymphatic system. It may either start in, or be spread to, the lungs.

Symptoms of lung cancer vary, but often include a cough, chest pain, shortness of breath, wheezing, recurring lung infections, bloody or rust colored sputum, and hoarseness. Some tumors can press on large blood vessels near the lung and cause swelling of the neck or face. Other tumors press on certain nerves near the lung causing pain and weakness in the shoulder, arm or hand. Lung cancer can result in fatigue, loss of appetite and weight, headache, pain in other parts of the body, and bone fractures.

> Symptoms vary depending on the type and location of cancer. Lung cancer can spread through the bloodstream to other organs, especially with small cell carcinoma, making early diagnosis difficult.

A chest X-ray can detect most lung tumors, although a microscopic exam of a tissue specimen is needed to make an accurate diagnosis. A CAT scan, sputum cytology, needle biopsy, bronchoscopy, mediastinoscopy and X-rays of other organs are all important in diagnosis. Treatment depends on the location, size and type of cancer. Most noncancerous tumors are removed surgically to prevent future blockage or problems. Surgery is also an option for cancers other than small cell carcinoma that haven't spread beyond the lung. Small cell carcinoma is usually treated with chemotherapy, coupled with radiation treatment; this will substantially prolong survival in some patients.

SUMMARY

- The respiratory system involves the complex exchange of gases. Air is inhaled through the mouth and nose and carried to the alveoli. In the alveoli, oxygen diffuses into blood and carbon dioxide diffuses out. This carbon dioxide is then carried back up the airways and exhaled. This process of gas exchange provides all cells of the body with the oxygen so essential for life, and eliminates carbon dioxide, the main end product of metabolism.
- Breathing is controlled by the respiratory center at the base of the brain. The diaphragm contracts, enlarging the chest cavity, then relaxes to passively move air out of the lungs.
- Asthma is a chronic condition that involves airways that shut down and make the movement of air in and out of the lungs very difficult. Although it can't be cured, there are a variety of medications available to relieve the symptoms of asthma.
- Emphysema involves a permanent breakdown of the structure of the airways. Alveoli may be also destroyed, narrowed, collapsed or over–inflated. These changes seriously decrease respiratory function.

- World Trade Center Pulmonary Syndrome is a relative newcomer in the arena of lung disease. After the 2001 attack on the World Trade Center, many people in the immediate vicinity suffered from coughing, sore throats, sneezing and eye irritation.
- Lung cancer usually starts in the lining of the bronchi, but it can begin anywhere in the respiratory system, or metastasize from other parts of the body. Symptoms vary, but often include a cough, and/or chest pain, shortness of breath, wheezing, recurring lung infections, bloody or rust-colored sputum, and hoarseness.

CHAPTER 17

RENAL SYSTEM

KEY TERMS

interstitial	anticoagulant	cystocele
proximal	distal	

All cells of every animal bathe in an internal sea of fluid with which they exchange nutrients and other molecules essential for life. Cells depend on the fact that the composition of this "interstitial fluid" is constant. Much of the job of maintaining the staus quo is handled by the renal system.

THE JOB OF THE RENAL SYSTEM

The composition of the interstitial fluid is held constant by exchanges with blood across the capillaries of the vascular system. In this way, blood delivers fuels to, and picks up wastes from, the interstitial fluid. The fuels and nutrients in the blood come mainly from the digestive system. But where do the wastes and by-products of metabolism go? It is the renal system that helps remove these unwanted compounds from the blood. The renal system is critical for maintaining a constant blood composition, and is also essential for maintaining the composition of the interstitial fluid. This process whereby multiple systems work together to maintain a constant internal environment, is called homeostasis.

Nephrons

The main organs of the renal system are the kidneys (See Figure 17.1). Each of the two kidneys found in all vertebrates is made up of more than a million tiny structures called *nephrons*. In the simplest sense, these nephrons are really just tiny filters (See Figure 17.2). Like miniature sieves, they filter fluid from the blood, leaving blood cells and larger molecules behind. This filtered fluid (the filtrate) then flows through specialized hollow tubes (the tubules) to the renal pelvis, and from there to the ureter and on to the bladder. On its way out of the kidney, however, any valuable compounds in the filtrate are reabsorbed and returned to the blood. At the same time, other unwanted compounds still in the blood are pumped out of the blood and secreted into the filtrate. By these three main processes of filtration, reabsorption, and secretion, the nephrons precisely regulate the composition of both the blood, and the internal environment as well. Now let's take a closer look at each of these three main processes.

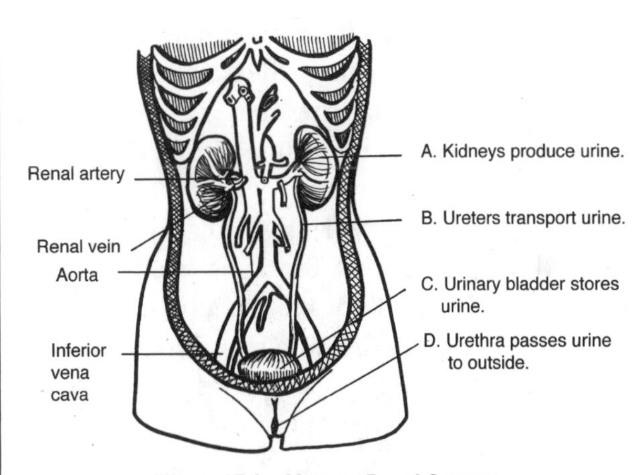

Figure 17.1—Human Renal System

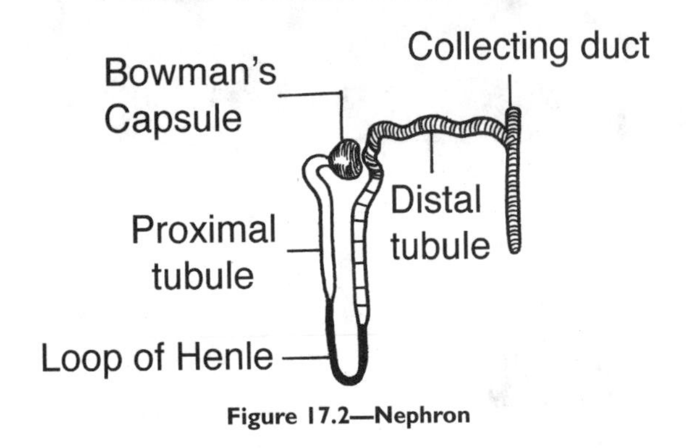

Figure 17.2—Nephron

COMPONENTS OF THE NEPHRON
Golmerulus—mechanically filters blood
Bowman's Capsule—captures the fluid filtered from the blood
Proximal Convoluted Tubule—reabsorbs seventy-five percent of the filtered water, salts, glucose and amino acids
Loop of Henle—establishes and maintains a salt gradient used to concentrate urine
Distal Convoluted Tubule—secretes hydrogen ions, potassium and certain drugs.

Filtration

Blood flows into the kidneys through the renal arteries. Inside the kidneys, the renal arteries give off numerous branches that get smaller and smaller. Most of these small branches supply blood to meshworks of capillaries that are very thin and leaky. This vascular structure is called the glomerulus. Blood pressure forces fluid out of the capillaries in the glomerulus, much like the water pressure in a garden hose forces water to leak from any holes in the hose. The fluid that is filtered out of the glomerulus is captured by a specialized structure that surrounds the glomerulus, called Bowman's capsule (together, a Bowmans' capsule and the glomerulus it surrounds are called a renal corpuscle). From the Bowman's capsule, the filtrate then flows on into the renal tubule.

Reabsorption

The first part of the tubule right after Bowman's capsule is the proximal tubule. In this section of the nephron, most water, glucose, and salts are reabsorbed into tiny capillaries (the peritubular capillaries) that surround the tubules. The process of reabsorption depends on the pumping of sodium out of the tubule. This pumping requires energy in the form of ATP, and is thus called active transport. It involves the diffusion of sodium in the tubular filtrate into the cells that make up the walls of the tubule. On the outside surface of these cells, sodium is actively pumped out of the tubule's cells, after which it diffuses into the peritubular capillaries. Because sodium in solution is a positively charged ion, when it is pumped it electrically attracts negatively charged ions like chloride to follow. This movement of ions out of the filtrate dilutes it and increases its water concentration. As a result, water then passively diffuses out of the tubule by osmosis. A passive process called facilitated diffusion also reabsorbs other molecules such as glucose. In this process, specialized carrier proteins help shuttle the glucose across the membranes of the tubular cells.

From the proximal tubule, the filtrate flows into a specialized part of the tubule called the Loop of Henle. This unique structure is essential for concentrating urine. The descending limb of the Loop of Henle is permeable to water, which flows out of the loop by osmosis. This removal of water from the filtrate concentrates it. At the end of the loop, the highly concentrated filtrate starts moving up the ascending limb of the Loop of Henle. In the ascending limb, salt is pumped out of the tubule and this concentrates the fluid surrounding the tubule. At the same time, however, the ascending limb is not permeable to water, so water remains within the tubule and dilutes the remaining tubular filtrate. The final result is that by the time the filtrate reaches the end of the Loop of Henle it is about the same concentration that it was at the beginning. The big differences are that there is a lot less filtrate left in the tubule, and there is also a big concentration gradient along the Loop of Henle. This concentration gradient is used later for further concentrating the urine.

After the Loop of Henle, the next part of the nephron is called the distal tubule. In this part of the tubule, more salt and water are reabsorbed and again, the active transport of sodium is involved. In the distal tubule, however, the rate of sodium pumping is precisely controlled by the hormone aldosterone. This hormone is released from the adrenal glands in response to signals from the brain that indicate when the body needs to conserve salt and water.

- **The brain releases a signal molecule called antidiuretic hormone (ADH) that helps to conserve water.**

ADH affects the collecting ducts, which are the part of the nephron that follows the distal tubules. The collecting ducts pass right by the ends of the loop of Henle where the salt concentrations are very high. The water permeability of the collecting ducts is very low, so most of the remaining filtrate in the collecting ducts flows on to the renal pelvis where it is collected and sent on to the ureter. However, in the presence of ADH, the water permeability of the collecting ducts increases dramatically. In this situation, water can then flow out of the collecting ducts into the regions surrounding the loop of Henle. This removal of water from the filtrate concentrates the urine. The collecting ducts are the last part of the nephron and are the last structure that can change the composition of the filtrate. After the filtrate leaves the collecting ducts, it is called urine and its makeup doesn't change further.

Secretion

Many unwanted compounds in blood are too large to be freely filtered in the glomerulus. Although the liver works to break down many of these undesirable compounds into small water–soluble molecules that can be filtered at the glomerulus, some products simply have to be secreted from the blood into the tubular filtrate. The process of secretion often involves active transport and thus requires ATP.

One of the main molecules secreted into the filtrate is hydrogen ion, the active component of all acids. The tubular membranes are not permeable to hydrogen ion so once it is in the filtrate, it stays there. Hydrogen ion can be formed, however, by the combination of water and carbon dioxide. The combination of these molecules is enabled by the enzyme carbonic anhydrase, which is on the inside of the tubules.

The combination of water and carbon dioxide produces a compound called carbonic acid, which quickly splits off into hydrogen ion and bicarbonate ion. The hydrogen ion remains trapped in the tubule and the bicarbonate ion is transported out of the tubule and into the blood where it can combine with another molecule of hydrogen ion and reduce the blood acid level. This process is carefully regulated and is influenced by how much and where the enzyme carbonic anhydrase is located, as well as by the amount of carbon dioxide in the blood. Still, the basic process is pretty simple. The kidney traps hydrogen ions in the tubule and reabsorbs the bicarbonate. In this way, hydrogen ion is secreted into the tubular filtrate.

Another ion that is secreted into the filtrate is ammonium ion. This is produced when a molecule of ammonia combines with a hydrogen ion in the tubular filtrate. Ammonia can move freely into and out of the tubule because it is not charged. When it combines with hydrogen ion, however, it becomes strongly charged and thus is trapped in the filtrate.

Potassium ion is also secreted into the tubular filtrate. Because most active transport of sodium involves movements in opposite directions by sodium and potassium, whenever sodium is moved out, potassium is left behind. In the proximal tubule, most potassium is reabsorbed by diffusion. In the distal tubule, potassium can be actively secreted into the tubular filtrate in direct exchange for sodium. The extent of this exchange is under the control of the hormone aldosterone, which, as mentioned earlier, is released from the adrenal glands in response to regulatory signals from the brain.

Many other small molecules are also secreted from the blood into the tubular filtrate. These include creatinine (a by-product of muscle metabolism), drugs like penicillin, and many other compounds. Most of these are transported actively in both the proximal and distal tubules, as well as in the collecting ducts.

Excretion

From the collecting ducts, the urine flows into the renal pelvis, which is a collecting area in the center of each kidney. From there, the urine flows out of the kidneys through the ureters and into the bladder where it is stored.

- **The movement of urine through the ureters is helped along by tiny waves of contractions called ureteral peristalsis.**

From the bladder, the urine flows to the outside of the body through a tubular structure called the urethra. Rings of muscle around the junction of the bladder and urethra help to contain the urine. When the bladder is full, special sensors send signals to the brain. At the right time, the brain allows the emptying of the bladder through a special reflex called the micturition reflex. In this reflex, a full bladder sends signals through the spinal cord which come back to signal the bladder to contract and also signal the sphincters around the urethra to relax. At the same time, sphincters where the ureters meet the bladder tighten, so that urine is not forced back into the kidneys.

Most people urinate about four to six times a day during waking hours, but this frequency can be changed dramatically by many conditions, the simplest of which is drinking too many fluids late at night (especially alcohol, coffee or tea). The most common cause of increased frequency of urination is urinary tract infections. Other causes of frequent urination include diseases of the heart and liver as well as diabetes.

Overall, our nephrons produce about 125 ml of filtrate each minute. At this rapid rate, the body's entire fluid content is filtered about sixteen times a day, which totals about

180 liters of filtrate each day. Of this, all but about 1.5 liters are reabsorbed and this remainder is excreted as urine. When everything is working well, we hardly notice how beautifully our kidneys perform. But when anything starts to go wrong, we notice right away.

> The kidney secretes two hormones. Renin is involved in salt and water balance and so helps control blood pressure. Erythropoietin stimulates red blood cell production by the bone marrow.

KIDNEY DISORDERS

Kidney disorders are diagnosed several ways including chemical analysis of the urine, which can detect protein in the urine (proteinuria), the presence of sugars or ketones, or even blood cells. Protein in the urine indicates breakdown of glomerular membranes and is a sign of serious kidney disease. Sugar or ketones in the urine signals the possible presence of diabetes. White blood cells in the urine suggest a possible kidney infection. Tissue and cell sampling via kidney biopsy are also quite useful in detecting cancers and monitoring the progress of treatment.

Many causes, such as infection, toxins, and genetic disease can produce acute kidney failure. Acute renal failure is simply defined as a rapid decline in the kidneys' ability to clear the blood of toxic substances, which leads to an accumulation of metabolic waste products, such as urea, in the blood. Acute kidney failure can indicate at least three separate problems:

- insufficient blood supply, dehydration or physical injury, heart failure, or liver failure
- obstructed urine flow due to an enlarged prostate or a tumor pressing on the urinary tract
- injuries within the kidneys that might include allergic reactions, toxic substances, blocked arteries, and crystals, protein or other substances in the kidneys.

Acute kidney failure can often be treated successfully. If kidney failure is severe, then dialysis is indicated, sometimes indefinitely or until a transplant is available. Chronic kidney failure is a slower, progressive decline in kidney function. Causes can include high blood pressure, urinary tract obstruction, kidney abnormalities, such as polycystic kidney disease, diabetes and autoimmune disorders. Many of the symptoms can be controlled by diet, water intake and medications. Long-term dialysis or transplantation are options only when other treatments are no longer effective.

Dialysis

Both hemodialysis and peritoneal dialysis remove waste products and excess water from the body. Hemodialysis is the process of removing blood from the body, pumping it into a machine that removes toxic substances by diffusion, and then returns the purified blood to the patient. Peritoneal dialysis requires that fluid containing a special mixture of glucose and salts is infused into the abdominal cavity where it draws toxic substances from the tissues. The fluid is then drained and discarded.

Dialysis is usually done about three times a week, depending on the severity of kidney disease, and allows patients to lead fairly normal lives. Special diets and drugs are important, and the cost, in terms of treatment and medications, loss of independence and the patient's time have a significant impact on the patient's quality of life, even though dialysis keeps them alive.

- **Nephritis is an inflammation of the kidneys that can be caused by an infection or an abnormal immune reaction.**

BLOOD VESSEL DISORDERS

To function properly, the kidneys require a continuous blood supply. Any disruption of blood flow can cause damage. Total blockage of a renal artery is rare, but can come from large particles in the bloodstream, a blood clot, or an injury caused by surgery.

Small blockages don't cause many symptoms, except for perhaps a steady aching pain in the lower back. Total blockage of both renal arteries completely stops urine production and shuts down the kidneys. Imaging of the kidney is the only way to make an accurate diagnosis of these kinds of problems.

Blood vessel disorders can include inflammation of blood vessels, blockage of the renal artery, blockage of small kidney blood vessels, damage to all or part of the outer layer of the kidneys, damage to small blood vessels in the kidneys caused by high blood pressure, and blockage of the renal vein. Typical treatments are anticoagulants, and sometimes surgery.

Some kidney abnormalities can be anatomic or metabolic and often are hereditary and present at birth. These include Renal Glycosuria, Nephrogenic Diabetes Inspidus, cystinuria, and other, more rare hereditary diseases.

Urinary Tract Infections, Obstructions, and Incontinence

Aside from diseases that directly affect the kidneys, the lower urinary tract is subject to a host of simple but annoying afflictions. Urine in the bladder is generally sterile, but any part of the urinary tract can become infected. Such infections often begin in the urethra where it is called urethritis, but may move to the bladder where it is called cystitis. If left untreated it may move up the ureters (urteritis), to the kidneys (pyelonephritis) where it can be very dangerous and cause severe kidney damage. These infections can be caused by bacteria, viruses, fungi, and parasites.

A slow or hesitant start when urinating can indicate an obstructed uretha in women (perhaps due to a growth or swelling) or an enlarged prostate in men. Urinary tract obstructions can occur anywhere along the urinary tract, from the kidney to the urethra. Common ailments include hydronephrosis, or distention of the kidney, and stones in the urinary tract.

Urinary incontinence is the uncontrollable loss of urine and this simple disorder impacts great numbers of people of all ages. Women are more likely to be affected, and 50 percent of all nursing home residents are incontinent. In women, incontinence is often caused by a cystocele, which is the result of stretching and weakening of the pelvic muscles during childbirth. Female urinary incontinence is also more prevalent after the decreases in estrogen that occur after menopause.

- **Many people live with incontinence when it could be cured or controlled if treatment is started early.**

Injury to the kidneys or urinary tract can result from many things including surgery, radiation therapy, blunt trauma or penetrating wounds. Treatment depends on the type of injury, and what the symptoms might be for the patient.

Kidney and Urinary Tract Tumors and Cancers

Bladder cancer is the leading type of new cancers in men, and for women ranks eighth in new cancer diagnosis. Cancer of the kidney accounts for about two percent of cancers in adults. Urine may be visibly bloody when bladder cancer occurs, and a routine microscopic examination of a urine specimen can detect red blood cells.

Blood in the urine is also the first symptom of adenocarcinoma of the kidney, but sometimes this isn't noticeable. Other symptoms might include pain in the side and fever. If the cancer hasn't spread to other organs, surgical removal of the affected kidney and lymph nodes can be successful. If it has spread to other sites (frequently the lungs), then treatment with interleukin-2 has proven helpful.

Cancer can also occur in the renal pelvis, ureters, and urethra. Blood in the urine, or pain in the lower abdomen are early indications of cancer of the renal pelvis and ureter. Cancer of the urethra is much more rare, but again one of the first symptoms is blood in the urine. A biopsy is necessary for diagnosis and surgical removal of the cancer is often successful.

SUMMARY

- The renal system consists of the kidneys, ureters, bladder and urethra. The main function of this system is to regulate body fluid composition.
- The kidneys filter metabolic waste, excess sodium and water from the blood, and also regulate blood pressure and red blood cell production. Each kidney contains about a million filtering units, or nephrons.
- Blood flows into the kidney through the renal artery. Water, sodium glucose and other filtered substances are reabsorbed into the bloodstream, while other unwanted molecules flow toward the ureters, then into the bladder. The bladder contracts and expels urine through the uretha.
- Diseases of the renal system can range from a relatively simple urinary tract infection to acute or chronic kidney failure.
- Other problems in the kidneys include infection, the inability of the kidneys to discard waste products, injuries and blood vessel disorders. All of these can be treated in a variety of ways, depending on the seriousness of the problem. In chronic kidney disease, dialysis is often a treatment that allows a patient to lead a full, relatively normal life.
- Cancers of the renal system are not unusual. Blood in the urine is often a first indication of something wrong. Cancer can occur in any part of the system and, depending on the severity and the extent, surgery and drug therapies can prove successful.

CHAPTER 18

ENDOCRINE SYSTEM

KEY TERMS

secretion	myocyte	cholesterol
glycoproteins	pregnenolone	circadian
catecholamine		

One of the miracles of the human body is that in spite of the fact that we don't always eat regular meals or maintain constant blood concentrations of glucose, our body manages to adjust quite nicely. Chemical messengers called hormones make up for many of our poor choices by stimulating reactions that result in glucose concentrations back in the normal range. The endocrine system produces a wide variety of hormones for this job and many others. In fact, the list of chores that fall to the endocrine system is a very long one.

THE ENDOCRINE GLANDS

The endocrine system consists of a group of organs, often referred to as glands of internal secretion. These glands broadcast hormonal messages to all cells by secretion into blood and extracellular fluid. The nervous system often acts with the endocrine system to regulate physiology.

The major organs of the endocrine system are the hypothalamus, the pituitary gland, the thyroid gland, the parathyroid glands, the islets of Langerhans of the pancreas, the adrenal glands, the testes, the ovaries and the pineal. Along with the classic endocrine organs, many other cells in the body secrete hormones. Myocytes in the atria of the heart, and scattered epithelial cells in the stomach and small intestine, are examples of the "diffuse" endocrine system (See Figure 18.1).

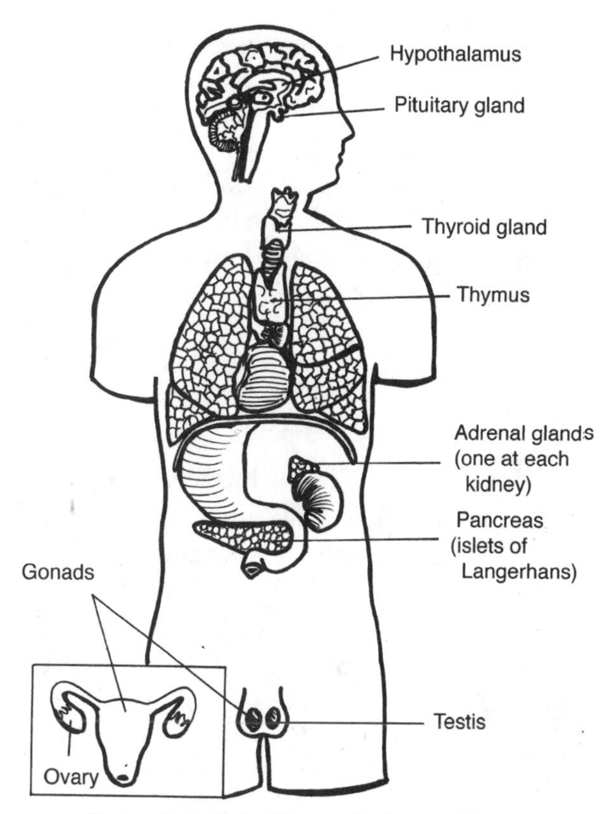

Figure 18.1—Major Human Endocrine Glands

The endocrine system and nervous system are functionally and structurally connected. The hypothalamus is the region of the brain that controls blood pressure, body temperature, and emotions. It is also an endocrine gland that produces hormones, and is physically connected to the pituitary gland. The pituitary gland is referred to as the master gland because it controls the functions of other endocrine glands, including the thyroid, adrenal cortex, ovaries, and testes. Some pituitary hormones have direct effects, such as growth hormone that stimulates tissues of the body to grow. Other hormones control the rate at which the other glands secrete hormones. For example, through adrenocorticotropic hormone, the pituitary stimulates the adrenal cortex to secrete steroid hormones that affect metabolism of fat, glucose, and protein.

The pituitary gland actually controls the rate at which it secretes its own hormones through a feedback loop. Levels of other endocrine hormones in the blood can signal the pituitary to slow down or speed up hormone release.

Types of Hormones

Hormones circulate in the blood, coming into contact with virtually all cells. The effect of the hormone on the cell depends upon the type of hormone, and the type of cell. Most hormones are proteins comprised of amino acid chains of varying lengths, or steroids, derived from cholesterol.

The role of hormones in the body is significant. Some (peptide receptors) bind to receptors on a cell. This triggers a cascade of reactions within the cell, speeding it up, slowing it down, or changing its function. Other hormones (steroid receptors) penetrate the cell membrane and act within the cell, usually to change gene expression.

Hormones can trigger responses in three ways:

- **Endocrine action**: the hormone moves via the blood and binds to distant target cells.
- **Paracrine action**: the hormone acts locally by diffusing from its source gland to nearby target cells.
- **Autocrine action**: the hormone acts on the same cell that produced it.

Hormones have been categorized into four structural groups, although members of each group have a lot in common. They are peptides and proteins, steroids, amino acid derivatives, and fatty acid derivatives, or eicosanoids (or fatty acid derivatives).

Peptides and Proteins

Peptide and protein hormones are products of translation, the formation of an amino chain from the genetic information provided by mRNA. They vary greatly in size and range from three amino acids to large, multisubunit glycoproteins.

Peptide hormones are synthesized in the endoplasmic reticulum, transferred to the Golgi and moved via secretory vesicles. Some cells can store peptide hormones in secretory granules, and release them in "bursts" when stimulated. This is most common, and allows cells to secrete a large amount of hormone over a short period of time. In constitutive secretion (made on demand) the cell does not store the hormone, but secretes it from secretory vesicles as it is synthesized.

Steroids

Steroid hormones are lipids, or more specifically, derivatives of cholesterol. Examples include testosterone, and adrenal steroids like cortisol which increase blood sugar. The first step in the synthesis of all steroid hormones is the conversion of cholesterol to pregnenolone. Pregnenolone is formed on the inner membrane of mitochondria, then shuttled back and forth between mitochondrion and the endoplasmic reticulum for further enzymatic transformations.

Newly synthesized steroid hormones are rapidly secreted from the cell, with little, if any, storage. Increases in secretion reflect accelerated rates of synthesis. Following secretion, all steroids bind to some extent to plasma proteins.

Amino Acid Derivatives

There are two groups of hormones derived from the amino acid, tyrosine. Thyroid hormones (such as throxine, which increase the rate of cell metabolism) are double tyrosine with the critical incorporation of three or four iodine atoms. The circulating half-life of thyroid hormones is a few days. Catecholamines, which include epinephrine and norepinephrine (see page 147), function as both hormones and neurotransmitters. They degrade rapidly, circulating for only a few minutes. Two other amino acids used for synthesis of hormones are tryptophan (a precursor to serotonin and the pineal hormone melatonin), and glutamic acid which is converted to histamine.

Fatty Acid Derivatives

Fatty acid derivatives, or eicosanoids, are a large group of hormones derived from polyunsaturated fatty acids (any acid derived from fat by hydrolosis). The eicosanoids are secreted by all cells, with the exception of red blood cells. The main groups are prostaglandins, leukotrienes, and thomboxanes. Arachadonic acid is the most abundant precursor for these hormones. Stores of arachadonic acid are present in membrane lipids and released through the action of various lipases (fat-splitting enzymes). The processing enzymes present and active in a cell dictate the specific eicosanoids synthesized. These hormones are typically active for only a few seconds.

MAJOR ENDOCRINE GLANDS

Pituitary and Hypothalamus

The hypothalamus makes hormones that control the pituitary gland. Examples are corticotropic RH (stimulates release of ACTH) and gonadotropic RH (stimulates release of LH and FSH). It also makes the antidiuretic hormone (ADH) and oxytocin (OT), which are stored in the pituitary. This grape-sized gland is at the base of the brain, just under the hypothalamus.

The pituitary has two separate parts, the anterior (front) and the posterior (back) lobes. The hypothalamus controls the anterior lobe by releasing control hormones in the blood, and it controls the posterior lobe through nerve impulses (See Figure 18.2).

The anterior lobe hormones include:

- **growth hormone** (GH), which stimulates the division of growth of bone, muscle and other body cells
- **thyroid stimulating hormone** (TSH), which stimulates the development and activity of the thyroid gland
- **adrenocorticotropic hormone** (ACTH), which stimulates the adrenal glands to secrete other hormones
- **follicle-stimulating hormone** (FSH), which stimulates the gonads to produce eggs and sperm

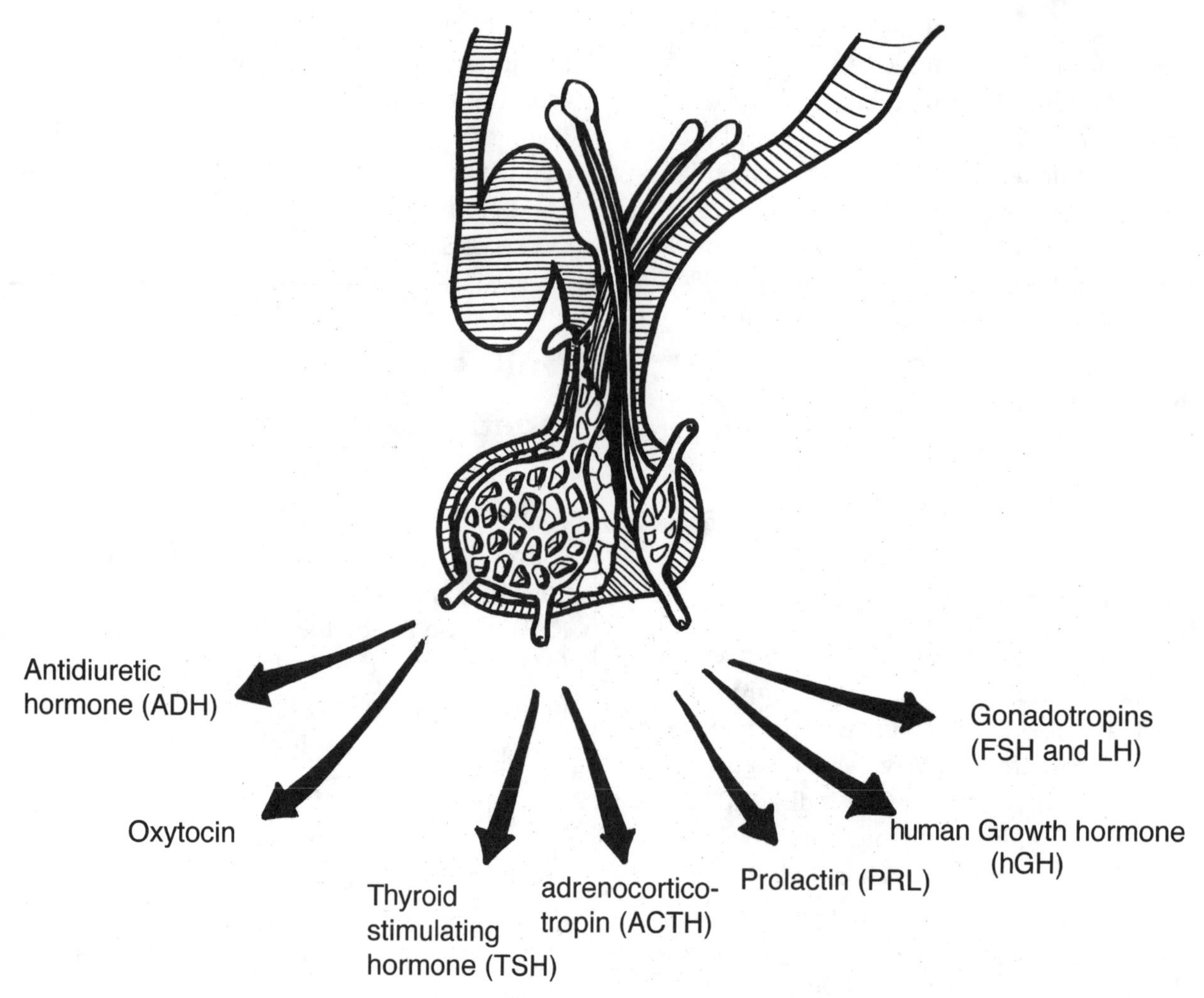

Figure 18.2—Hypothalamus and Pituitary

- **luteinizing hormone** (LH), which stimulates the gonads to produce sex hormones.
- **prolactin** (PRL), which stimulates milk prodcution.
- **human growth hormone** (hGH), which stimulates growth and fat mobilization.

The posterior lobe hormones include:

- **antidiuretic hormone** (ADH), which controls water absorption by the kidney
- **oxytocin,** which stimulates uterine contractions during childbirth and the flow of milk while nursing.

A function of the hypothalamus and the pituitary gland is to regulate the reproductive system by stimulating the sex organs to secrete sex hormones: testosterone for males, estrogen and progesterone for females.

Not all hormones produced by the pituitary and hypothalamus are released continuously. Some, such as corticotropin, growth hormone, and prolactin follow a circadian rhythm (biologic rhythms with a cycle of about twenty-four hours). The timing of the release of other hormones varies.

The pineal gland, also located adjacent to the hypothalamus, secretes melatonin, which controls body functions in response to daylight and seasonal changes.

Thyroid and Parathyroid

The thyroid, almost two inches across, lies just under the Adam's apple in the neck. It has two lobes connected in the middle by the isthmus. The thyroid secretes the hormone thyroxine which speeds up metabolism and helps manage growth and development. Thyroid hormones influence the metabolic rate by stimulating almost every tissue in the body to produce proteins, and by increasing the amount of oxygen that cells use. When the cells work harder, the body responds in like fashion.

To function properly the thyroid gland must trap iodine, an element contained in our food and water, and then process this iodine into thyroid hormones. As the hormones are used up, some iodine returns to the thyroid gland and is recycled there.

In an amazing sequence of events, a complex system allows the thyroid to work. The hypothalamus secretes thyrotropin-releasing hormone (TRH), which in turn causes the pituitary to produce thyroid stimulating hormone (TSH). The pituitary responds to the level of thyroid hormones circulating in the blood, cutting back or picking up production as the body needs.

About eighty percent of thyroxine is converted in the liver and other organs to a metabolically active form, triiodothyronine. The rest is converted in the thyroid. The conversion rate of triiodothyronine relates directly to the body's needs at the moment, and remains tightly bound to proteins in the blood until it is required.

- **The parathyroids are four patches of tissue on the thyroid gland that release the parathyroid hormone, a regulator for blood calcium level.**

The thymus gland, located in the upper chest, secretes thymosin, which stimulates the development of T cells for the immune system. It is an important part of the lymphatic system, defending the body against infection.

Adrenal

The adrenal glands make epinephrine and norepinephrine, two hormones which cause the "fight or flight" response. They also secrete aldosterone, which affects the body's salt and water balance, and cortisol, which promotes glucose synthesis and facilitates responses to stress.

The adrenal glands are each located near the top of a kidney. The inner part of the adrenal gland (medulla) secretes adrenaline (epinephrine) which affects blood pressure, heart rate, sweating, and other activities also regulated by the sympathetic nervous systems. The outer part (cortex) secretes, among other things, corticosteroids, androgens, and mineralocorticoids, which help to control blood pressure and the levels of salt and potassium in the body.

The adrenal glands are at the mercy of a properly functioning hypothalamus and pituitary. The adrenal glands may stop working if sufficient amounts of other proteins are not produced first.

Pancreas

The pancreas contains two basic types of tissue, the acini, which produce digestive enzymes, and the islets of Langerhans, which have cells that make the hormones insulin and glucagon. Insulin and glucagon control the blood sugar level. Glucagon raises the level of sugar in the blood, insulin lowers the level of sugar, and somatostain, also produced by the pancreas, prevents the first two from being released.

The pancreas also secretes digestive enzymes into the duodenum (first division of the small intestine), and hormones into the bloodstream. Digestive enzymes are released from the cells of the acini and flow down various channels into the pancreatic duct. It joins the common bile duct at the sphincter of Oddi, where both flow into the duodenum. These enzymes are important in the digestion of proteins, carbohydrates, and fats. These enzymes are only activated when they reach the digestive tract.

Ovaries and Testes

The ovaries in females, and the testes in males, are part of the endocrine system. Estrogen and progesterone are made in the ovaries, and maintain the female reproductive system and secondary sex characteristics. Progesterone maintains the uterus during pregnancy. The testes make testosterone, a hormone that maintains the male reproductive system and secondary male sex characteristics.

ENDOCRINE SYSTEM DISORDERS

The endocrine system functions well as a unit. When one part is not working, it can have a serious effect on the entire body. Each message carried by the system must be sent and received properly to keep everything working correctly.

One of the most serious endocrine disorders is diabetes mellitus, in which the body is unable to control blood sugar levels. A dangerously high level of glucose in the blood can result in limb loss, blindness, and even coma or death. Sugar in the urine is one of the significant warning signs of diabetes because the person with this disorder releases large amounts of glucose in the urine.

Type I diabetes, usually develops during childhood. The beta cells of the islets of Langerhans in the pancreas do not produce insulin. Patients need regular injections of insulin, and a carefully controlled diet.

Type II diabetes usually develops in people over the age of forty. In these cases, the beta cells may produce insulin, but body cells do not respond to the hormone properly. Type II diabetes is controlled with exercise, medication and diet.

Other disorders of the endocrine glands usually affect growth or metabolism. One condition, hypothyroidism, causes the body's metabolism to slow down. The patient feels sluggish and cold, and may gain weight easily. In hyperthyroidism a person feels irritable, has high body temperature, sweating, weight loss, and high blood pressure.

Stress can also have an impact on the functioning of the endocrine system. Short-term stress causes the adrenal glands to secrete epinephrine and norepinephrine, providing bursts of energy to deal with a crisis. This is usually accompanied by rising blood pressure and blood glucose levels. Long-term stress can cause the adrenal glands to secrete excessive amounts of the hormone cortisol, which can lead to high blood pressure and a suppressed immune system.

SUMMARY

- The endocrine system involves the secretion and release of hormones throughout the body. Hormones contribute to the control of organ function, growth and development, reproduction and sexual characteristics, how the body uses and stores energy, and the volume of fluids and levels of salt and sugar in the blood.
- Major organs of the endocrine system are the hypothalamus, the pituitary gland, the thyroid gland, the parathyroid glands, the islets of the pancreas, the adrenal glands, the testes and the ovaries.
- The pituitary produces hormones that control many functions including water absorption by the kidney. The pineal gland, located adjacent to the hypothalamus, secretes melatonin.
- The thyroid secretes the hormones that speed up the metabolism and manage growth and development. The parathyroids are four patches of tissue on the thyroid gland that release a regulator for calcium level. The thymus gland stimulates the development of T cells for the immune system.
- The adrenal glands make epinephrine and norepinephrine, the two hormones which cause the "fight or flight" response. The pancreas produces digestive enzymes, and the hormones insulin and glucagon, which control the blood sugar level.
- The ovaries, in women, produce estrogen and progesterone, and the testes, in males, produces testosterone.
- A serious endocrine disorder is diabetes mellitus, when the body is unable to control blood sugar levels. Type I diabetes usually develops during childhood. Type II diabetes usually develops in people over forty.

CHAPTER 19 NERVOUS SYSTEM

KEY TERMS

ganglia	myelin sheath	autonomic
neurotransmitter	synapse	

The nervous system has two main parts, the central nervous system, which is the body's primary control center consisting of the brain and spinal cord, and the peripheral nervous system, a network of nerves that extend throughout the body. The two systems work together to monitor, coordinate, and control the activities of the entire body.

CENTRAL NERVOUS SYSTEM

The central nervous system processes information and sends instructions to other parts of the body via the peripheral nervous system. The peripheral nervous system gathers information and returns it to the central nervous system.

The Brain

Our highly-developed brain makes us distinctly human. It is the basis of our emotions, memories, behaviors and moods. Because of our brain we can read and write, compose and appreciate music, communicate with others, and plan for the future. The brain is "command central" for the nervous system.

The brain is a complex, quick-acting miracle. It requires constant and plentiful nourishment to keep it working at its optimum. It uses about twenty percent of the blood flow from the heart. Lack of oxygen, abnormally low blood sugar levels, or toxic substances cause the brain to malfunction within seconds. Fortunately, the body has mechanisms to help prevent this from happening.

- **An interruption of blood flow to the brain for even ten seconds can cause loss of consciousness.**

The brain weighs about 1500 grams, and contains approximately 100 billion neurons, and 900 billion glial cells (cells that protect and nourish neurons). It is protected by bones, tough membranes called meninges, and a cushion of fluid. It has three main elements, the cerebrum, the cerebellum, and the brain stem (See Figure 19.1).

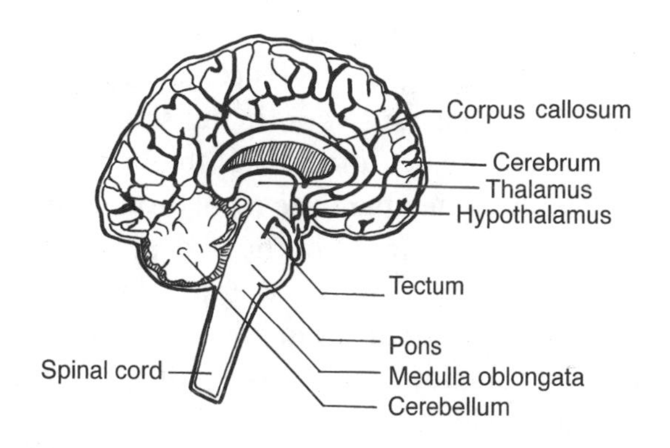

Figure 19.1—Human Brain

Cerebrum and Thalamus

The cerebrum is responsible for all voluntary, or conscious, activity in the brain. It is covered with dense, convoluted masses of tissue divided into two halves, the left and right cerebral hemispheres, connected in the middle by nerve fibers called the corpus callosum. These hemispheres are divided into four lobes, the frontal, parietal, occipital, and temporal lobes. Each hemisphere controls activities of the opposite side of the body.

- **The frontal lobe** controls skilled motor behavior including speech, mood, thought, and planning for the future.
- **The parietal lobe** interprets sensory input from the rest of the body and controls body movement.
- **The occipital lobe** interprets vision.
- **The temporal lobe** generates memory and emotions. It allows people to recognize other people and objects, process and retrieve long-term memories, and initiate communication or action.

At the base of the cerebrum are the basal ganglia, thalamus, and hypothalamus. The basal ganglia help regulate movement. The thalamus organizes sensory messages passing to and from the highest levels of the brain, and the hypothalamus coordinates some of the more automatic functions of the body, like sleep, body temperature regulation, and control of water balance in the body. The brain stem, which connects the spinal cord to the rest of the brain, is involved in many important functions such as breathing, heart rate control, sleep, and wakefulness. It is divided up into three regions: the midbrain, the pons, and the medulla oblongata.

The cerebellum, located above and behind the brain stem, is the second largest part of the brain and it coordinates muscle activities, enabling the body to move in a smooth, steady and efficient manner. It is also in charge of balance, equilibrium, and posture.

Spinal Cord

The spinal cord is a tube-like, fragile organ, consisting of neurons, supporting tissue, and blood vessels. It is protected by the bones of the spine, three membranes called the meninges, and a cushion of fluid.

The job of the spinal cord is to relay nerve impulses to and from the brain. The brain communicates with much of the body through nerves that run up and down the spinal cord. Each vertebra has an opening between it and the one above and below it; through this opening, spinal nerves branch out and carry messages to other parts of the body.

ORGANIZATION OF THE HUMAN NERVOUS SYSTEM

The body's nervous systems consists of two components:

I. Central nervous system (encased in the skull and spine)

II. Peripheral nervous system:

   A. Sensory nerves (monitor internal body conditions and external surroundings)

   B. Motor nerves (control moment)

      1. Somatic nervous system—skeletal muscle

      2. Autonomic nervous system—smooth muscle, cardiac muscle, and glands

         a. Sympathetic nervous system (generally excites the body)

         b. Parasympathetic nervous system (generally relaxes the body)

Motor nerves at the front of the spinal cord carry messages from the brain to the muscles. The sensory nerves, at the back of the spinal cord, carry sensory information from distant parts of the body to the brain.

The spinal cord's outer region is white matter and the inner region is gray matter. Neurons in the white matter are protected by cells called oligodendrocytes. They form a sheath that surrounds neurons, which is why this matter looks white. The internal neurons don't have such a sheath, making them appear gray.

PERIPHERAL NERVOUS SYSTEM

The peripheral nervous system is comprised of bundles of single nerve fibers, some less than one millimeter in diameter and others more than six millimeters in diameter. The *somatic nervous system* is part of the PNS and controls voluntary responses. Skeletal muscles respond to the somatic nervous system, and it is responsible for some involuntary reflexes as well.

Nerves that communicate between the brain stem and the body's internal organs are. called the *autonomic nervous system*, which regulates internal body processes that require no conscious awareness, such as the rate of heart contractions, breathing, stomach acid secretion, and the speed at which food passes through the digestive tract.

The autonomic nervous system is comprised of two parts, the *sympathetic nervous system,* which responds to stressful situations, and the *parasympathetic nervous system*, which controls body functions associated with rest and digestion.

The nerves of these two systems often have opposite effects on organs and glands. Stimulating a sympathetic nerve speeds heart rate. Stimulating a parasympathetic nerve slows the heart down. The opposing effects help the body maintain its balance.

Parasympathetic nervous system:

- contracts pupils
- stimulates salivation
- contracts bronchi
- slows heartbeat
- stimulates digestive activity
- stimulates gallbladder
- contracts bladder

Sympathetic nervous system

- dilates pupils
- inhibits salivation
- relaxes bronchi
- accelerates heartbeat
- inhibits digestive activity
- stimulates glucose release by liver
- relaxes bladder

NEURONS

Nerves carry thousands of messages throughout your body every moment of the day. Nerve impulses are waves of chemical and electrical changes that move along the membrane of a neuron. These changes can take place in a few thousandths of a second.

There are 100 billion or more nerve cells running throughout the body. Each nerve cell, or neuron, has a large cell body, and an elongated extension (axon), which delivers chemical messages (See Figure 19.2). Neurons

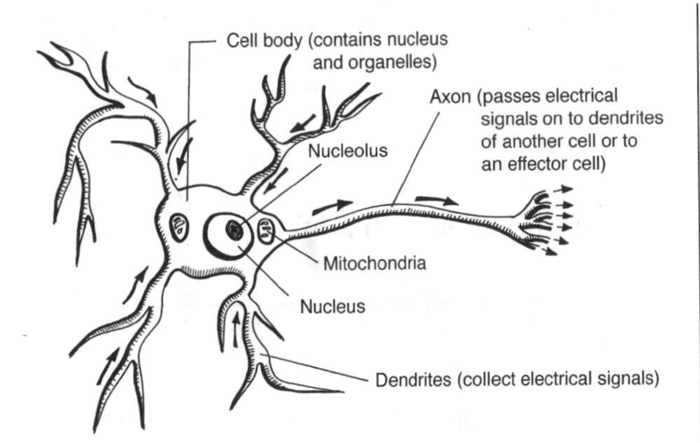

Figure 19.2—Structure of a Neuron

also have many dendrites, or branches, that receive messages. Simply put, nerve signals come in on the dendrites, and go out on the axons.

The neuron has an electrical charge across the cell membrane. This is due to the pumping of charged ions across the membrane. The inside has a negative charge relative to the outside. When fully stimulated the charges across the cell membrane reverse directions and then return to "normal" very quickly. This electrical event allows a nerve to pass on signals to other nerves and is called an action potential.

The neurons transmit their messages electrically in one direction, from the axon of one neuron to the dendrite of the next. The transmission of a nerve impulse involves the movement of ions across a neural membrane. At synapses, or junctions with tiny gaps between adjacent neurons, the message sending axon secretes tiny amounts of chemicals called neurotransmitters. Channels in the membrane of the sending neuron open, and sodium ions (Na^+) flow into the cell. After Na^+ enters the neuron, potassium ions (K^+) flow out of the cells, restoring the cell's electrical charge to its resting potential. The neurotransmitters released by the sending neuron trigger receptors on the next neuron's dendrite to start up a new electrical current, causing the next neuron to fire. The charge reversals continue as the nerve impulse is transmitted from neuron to neuron.

Each axon is surrounded by an insulating myelin sheath. When a nerve impulse reaches the end of an axon, the impulse causes sacs containing neurotransmitter to fuse with the axon's membrane. Each fused sac releases its neurotransmitter into the synapse gap between adjacent neurons. It diffuses across the gap and binds to receptors on the next cell that is receiving the message. This binding causes a change in the membrane potential of the receiving cell, and then leads to a new nerve impulse or response.

Sensory neurons monitor body and environmental conditions and conduct nerve impulses toward the central nervous system. Motor neurons conduct impulses away from the central nervous system and cause muscles, organs and glands to respond.

- **Interneurons are special neurons that carry messages from sensory to motor neurons (and vice versa).**

NERVOUS SYSTEM DISORDERS

Billions of neurons acting at lightning speed, send important messages to organs, muscles, and glands. Sensory neurons help us respond and adapt quickly to our changing environment. This all happens without our even thinking about it, but when disease or injury attacks some part of the nervous system or the function of the neurons, the impact can be devastating.

When the myelin sheath is damaged, or interrupted, the information flow slows down or stops. This might be the result of Guillian-Barre syndrome or multiple sclerosis. Degenerating nerves might cause Alzheimer's disease or Parkinson's disease. Infections can cause meningitis or encephalitis, both of which can have a serious impact on the brain and nervous system functioning. Blood supply blockage might cause a stroke, and injuries or tumors can have a significant impact on the nervous system's ability to work.

Meningitis

Meningitis is a viral or bacterial infection in the brain compartment. Bacterial meningitis is the inflammation of the meninges, often by Neiseris meningitis, Hemophilus influenzae, and Streptococcus penumoniae. These pathogens are all in the environment, but rarely affect the brain. Some pre-existing conditions might predispose a person to meningitis.

- **Meninges are the membranous coverings of the brain and spinal cord.**

Symptoms include fever, severe headache, stiff neck, sore throat, and vomiting, followed by respiratory illness. Bacterial meningitis is swift in its course, making patients seriously ill within twenty-four hours. Diagnosis must be made quickly since untreated bacterial meningitis usually leads to death. Diagnosis is usually done by analyzing a sample of the fluid taken from around the spinal cord (a spinal tap). Antibiotics are the standard treatment. If not started soon enough, permanent brain damage can be the result, especially in young children or the elderly. Most young children are now immunized with Hemophilus influenzae type b, one of the most common causes of childhood meningitis.

Alzheimer's Disease

Alzheimer's is a form of dementia that affects a person's memory, ability to reason, make judgments and function in everyday situations. Although there is no definitive answer as to why some people get Alzheimer's disease, genetic factors are considered important.

In the Alzheimer's patient, brain cells degenerate, reducing the responsiveness of the remaining cells to many of the chemicals that transmit signals. Abnormal proteins and abnormal tissues, called senile plaques and neurofibrillary tangles, are often found at the time of autopsy. There are also lower levels of certain chemicals in the brain that help to carry complex messages through the body.

Symptoms of Alzheimer's disease are similar to dementia so it isn't always easy to tell, at first, whether or not it is specifically Alzheimer's. Changes can be subtle. The patient may not perform well at work, or there may be little memory lapses involving other aspects of a person's life. Sometimes the disease can begin with depression, fear, anxiety, decreased emotion, or other personality changes. The patient may begin to use simpler words, or have trouble finding the right word. Simple tasks become noticeably more difficult.

Alzheimer's affects up to four million Americans, and it is most common in people over sixty years of age. It is a slow disease and the course varies from person to person. Usually, a patient with Alzheimer's will live from eight to ten years after diagnosis.

There isn't a lot that can be done to stop the progression of Alzheimer's disease, but antipsychotic drugs, antidepressants, and a wide range of vitamins and nutritional supplements have proven helpful in lessening symptoms.

Parkinson's Disease

Parkinson's disease is a slow, degenerative disorder of the nervous system. Patients with this disease may exhibit excessive shaking, tremors, sluggish movements, or muscle rigidity.

The basal ganglia, located in the brain, are at the root of the problem in Parkinson's disease. The basal ganglia are responsible for processing signals and transmitting messages to the thalamus, which relays processed information to the cerebral cortex. These signals are transmitted by chemical neurotransmitters as electrical impulses along nerve pathways, and between nerves. The primary neurotransmitter is dopamine. When the nerve cells in the basal ganglia degenerate, there is less dopamine, and the connection between nerve cells and muscles is affected.

Exactly why the basal ganglia might degenerate isn't known, and although Parkinson's can occur in several individuals of the same family, it isn't thought to be genetic. Sometimes the cause can be viral encephalitis, ingestion of antipsychotic drugs, or drug abuse with a street-synthesized form of opiate known as N-MPTP. Head traumas and stroke can also be contributing factors.

Parkinson's might start as a simple hand tremor while the hand is at rest. It usually starts in one hand, progresses to the other hand, then the arms and legs. Not every Parkinson's patient has the tremor. It doesn't necessarily worsen as the disease progresses. In fact, the tremor may become less obvious.

Rigidity in muscles is one of the serious effects of Parkinson's. This rigidity and immobility can contribute to muscle ache and fatigue, and that, combined with hand tremors, make many small, daily tasks nearly impossible. As the disease progresses it becomes increasingly difficult to move around, leaving some patients confined to a wheelchair. There are a variety of drugs used to treat Parkinson's disease, but Levodopa-carbidopa is most common.

SUMMARY

- The human nervous system is the complex control center for the human body. The brain, spinal cord, and peripheral nervous system are constantly sending and receiving messages that help us to think, move, touch, see, smell, hear and communicate with others.
- The brain includes the cerebrum. The left and right cerebral hemispheres control activities on opposite sides of the body. The hemispheres are divided into four lobes responsible for different bodily functions, motor behavior, senses, vision, memory and emotions.
- At the base of the cerebrum are the basal ganglia, thalamus, and hypothalamus. The basal ganglia are basically responsible for coordinating body movements. The thalamus organizes sensory messages, and the hypothalamus coordinates automatic functions of the body.
- The cerebellum, behind the brain stem, coordinates muscle activities.

- The spinal cord is a tube-like, fragile organ. It is framed and protected by the backbone, the meninges, and a cushion of fluid. Along the spinal cord, billions of neurons exist within a myelin sheath.
- The peripheral nervous system consists of bundles of single nerve fibers. It regulates heart contractions, breathing, and other involuntary tasks.
- Neurons carry thousands of nerve impulses throughout the body. They transmit their messages through a series of reversal and restoration of electrical charges across the cell membranes.
- Some common nervous system disorders include meningitis, Alzheimer's disease or Parkinsons's disease. Meningitis, while it can be the result of a pre-existing condition, is usually caused by a viral or bacterial infection.
- Alzheimer's disease is a slow, progressive loss of the ability to remember, reason, think and function in everyday situations. Parkinson's disease is a degenerative disease, manifested first by hand tremors, and then increasing difficulties with muscle rigidity and sluggish movements.

CHAPTER 20 REPRODUCTIVE SYSTEM AND HUMAN DEVELOPMENT

KEY TERMS

sinus	endometrium	corpus luteum
blastocyst	zygote	gastrulation

One of the foundations of modern biology is the theory of evolution. Now try to imagine that process without sexual reproduction as a factor! Although some organisms survive asexually, the key to survival of most species, including humans, is an effective form of sexual reproduction.

MALE REPRODUCTIVE SYSTEM

The male's external reproductive system consists of the penis, scrotum, and testes. Internally, the male has the vas deferens, urethra, prostate gland, and seminal vesicles. The male has two testes, and at puberty they begin to produce large amounts of testosterone, signaling sperm cell production and the development of secondary sex characteristics.

The testes are inside the scrotum. Before birth, the male's testes are inside the pelvic cavity and they move down into the scrotum just before birth. The scrotum helps to protect the testes, and because it is outside the body, it keeps them cooler, an important element in sperm cell development and survival. The cremaster muscles in the scrotal wall relax to let the scrotum hang away from the body to cool, or move closer to keep warm.

The testes are made of seminiferous (semen-carrying) tubules that produce testosterone. Here, diploid cells undergo meiosis to produce haploid sperm cells. Each specialized diploid cell forms four equal-sized sperm cells. The sperm cell goes through many changes, first forming a long flagellus, or tail. Then the haploid nucleus condenses and becomes longer, forming a head. A tightly packed area of mitochondria forms between the head and tail. During this process, the sperm cell loses most of its volume and becomes more streamlined. The mature sperm are stored for a time in a coiled tube, almost twenty feet long, called the epididymis, which lies alongside each testicle inside the scrotum.

The root of the penis is attached to the middle of the abdominal wall, and the glans penis is

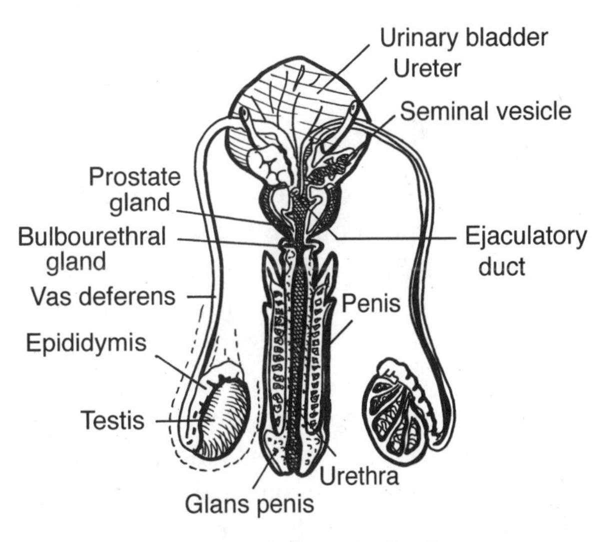

Figure 20.1—Male Reproductive Organs

the cone-shaped end (See Figure 20.1). The urethra runs through the length of the penis and ends at the tip. The base of the glans penis is called the corona.

The penis has three cylindrical spaces, or sinuses of erectile tissue. These corpora cavernosa run beside each other and the corpus sponglosum surrounds the uretha. When these sinuses fill with blood, the penis becomes large, rigid, and erect.

The vas deferens is a cordlike duct that transports the sperm from the epididymis, up through the prostate and into the urethra. Blood vessels and nerves also travel along the vas deferens, and together make the spermatic cord.

- **The urethra is both part of the urinary tract and part of the reproductive system. Through it, urine flows and semen is ejaculated.**

The prostate is just under the bladder and surrounds the middle portion of the urethra. The prostate and seminal vesicle above it produce fluid that nourishes the sperm, and provides most of the volume of the semen.

Prostate Cancer

Prostate cancer is extremely common in older men. It is a type of cancer that spreads very slowly and shows relatively few symptoms. Many men live it with for years, unaware of the problem.

When prostate cancer reaches its advanced stages, some of the symptoms include difficulty urinating and a frequent "need" for urination. This is usually because the cancer partially blocks the flow of urine. In its later stages, prostate cancer can cause bloody urine or sudden urinary retention. Prostate cancer itself may progress slowly and cause few symptoms, but if it metastasizes to the bone or the kidneys, the disease progresses rapidly, and the complications can have a much greater impact on a man's health.

Because prostate cancer is so common, screening is routine. This is done through a digital rectal exam and a special blood test for a marker called prostate specific antigen. The screening is not very reliable and there is some controversy about the value of having the exam and test, as well as about how and when to treat prostate cancer.

Surgery, radiation therapy and drugs often cause impotence, and sometimes incontinence. If the patient is over seventy, surgery may be contraindicated because the patient is much more likely to die of some other cause than prostate cancer. A newer treatment, radioactive seed implants, provide a less invasive approach. Tiny, radioactive pellets are implanted directly in the target tissue where they work to eradicate the cancer. Treatment must be chosen carefully to minimize life-long after effects; the choices often are based on the extent or size of the cancer, if it has metastasized, and the patient's age.

Testicular Cancer

Testicular cancer is characterized by a lump in the scrotum. The cause isn't known, but men whose testes failed to descend into the scrotum by age three have a higher incidence.

- **Unlike prostate cancer, testicular cancer seems to strike a much younger population, usually men under forty.**

The lump in the scrotum can be painful, and diagnosis is made quickly. Surgery, removal of the entire testis, is the usual course of treatment. Before surgery, a biopsy can be done to look for two proteins, alpha-feto-protein and human chorionic gonadotropin, which tend to be higher in men with testicular cancer.

Removal of one testis leaves a man with adequate levels of hormones and continued fertility. Often a combination of surgery and radiation will provide a good chance of recovery for men with testicular cancer. The outcome depends on the type of cancer—seminoma, teratoma, embryonal carcinoma, and choriocarcinoma—and whether it has spread to adjacent lymph nodes or other parts of the body.

FEMALE REPRODUCTIVE SYSTEM

The female reproductive system includes the ovaries, organs that release eggs and produce the hormones estrogen and progesterone; the uterus, which is an organ with strong muscular walls capable of protecting and nurturing a fetus; and the cervix, which connects the uterus with the vagina. The vagina is the birth canal and opening through which sperm can enter the body. Fallopian tubes carry eggs, which combined with the male sperm, may result in pregnancy.

The external genitalia include the vulva, which is bordered by the labia majora. These are similar to the scrotum in tissue origin, and contain sweat and sebaceous glands. The labia minora lie inside the labia majora, and surround the openings to the vagina and urethra. The labia minora meet at the clitoris, a small sensitive protrusion.

- **The opening to the vagina is called the introitus. Bartholin's glands beside the introitus secrete a fluid that supplies lubrication for intercourse.**

Internal genital organs include the vagina, the walls of which are normally touching except during intercourse. The vaginal cavity is three to four inches long. The cervix lies at the top of the vagina, connecting it to the uterus (See Figure 20.2).

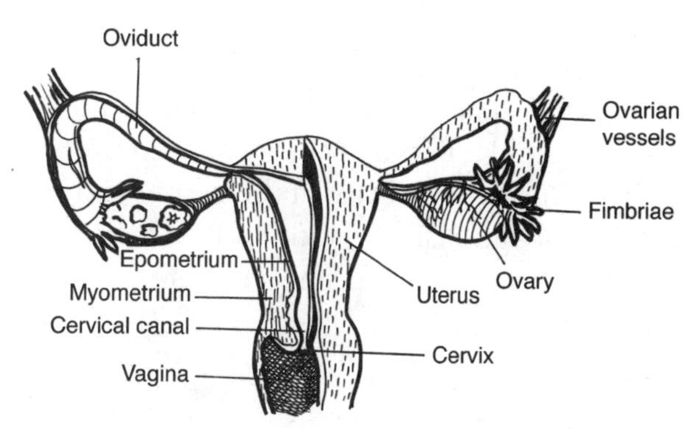

Figure 20.2—Female Reproductive Organs

The uterus is a pear-shaped organ that lies behind the bladder and in front of the rectum, and is held in place by six ligaments. Generally, the cervix provides a good barrier against bacteria. Unless a woman is ovulating, the mucus is thick and impenetrable to sperm. The consistency changes during the cycle to allow sperm to move through during ovulation, and menstrual discharges to exit after ovulation.

The fallopian tubes extend two to three inches from the upper edges of the uterus toward the ovaries. Muscles and cilia propel eggs released from the ovaries through the fallopian tube. If the egg is fertilized, it begins to divide. Within four days, as the tiny embryo continues to divide, it moves slowly down the fallopian tube and into the uterus. Implantation occurs when the fertilized egg secures itself to the wall of the uterus.

Hormonal Control of Menstrual and Ovarian Cycles

The *menstrual cycle* is the period during which an ovum (egg) matures, moves along the fallopian tubes, and enters the uterus. During the menstrual cycle, hormones from the ovaries prepare the endometrium (uterine lining), making it thicker. If the egg is fertilized, it can embed itself in the lining. If not, the lining is shed. This shedding marks day one of the menstrual cycle, which lasts about 28 days.

The *ovarian cycle* includes the development of an ovum-containing follicle, the rupture of the follicle, the release of the ovum, and formation and disintegration of the *corpus luteum*, a hormone-releasing structure.

FSH and LH

The hypothalamus stimulates the pituitary to release follicle-stimulating hormone (FSH) and later, luteinizing hormone (LH). FSH stimulates a follicle (structures in the ovaries where eggs develop). Lutenizing hormones cause a follicle to rupture and release its egg.

- **Female are born with many immature eggs called oocytes. At the first division of meiosis, the oocyte forms a mature egg, or ovum.**

During the follicular phase of the ovarian cycle, the oocyte becomes an ovum in a follicle. As it matures, follicle cells release estrogen, stimulating the endometrium to grow new tissue and blood vessels. This prepares the uterus to receive a fertilized egg. As the follicles produce increased amounts of estrogen, the hypothalamus triggers the pituitary gland to release increased amounts of LH and FSH. The increased LH triggers the follicle to release the ovum through a process called ovulation. The ruptured follicle becomes a corpus luteum. Thus begins the luteal phase of the ovarian cycle, from about day fifteen to the end of the cycle.

The corpus luteum releases large amounts of both progesterone and estrogen, which stops the pituitary from releasing LH and FSH, and causes the endometrium to grow even thicker.

The ovum moves into the fallopian tube. If fertilized, it will spend up to a week traveling to the uterus, where it will embed itself, resulting in pregnancy. If it is not fertilized, the corpus luteum will disintegrate. Levels of estrogen and progesterone and blood flow to the endometrium will decrease, resulting in menstruation. The disintegrating tissue and the unfertilized ovum leave the body.

Fertilization

Coordinated changes in gonadotropin, a hormone released by the anterior pituitary, egg development in the ovaries, the release of female sex steroids, and changes in the uterus must all happen at the right time for fertilization to occur.

- FSH and LH are controlled by GnRH (gonadotropin releasing hormone) from the hypothalamus and the ovarian hormones estrogen and progesterone.
- FSH stimulates the development of follicles; the LH surge causes ovulation and then the development of the corpus luteum.
- Ovarian hormones stimulate the development of the endometrium.

- Estrogen and progesterone control the development of the uterine lining.

The Female Reproductive Cyle

The cycle begins with:

1. bleeding and sloughing of uterine lining
2. low levels of estrogen and progesterone

Mid cycle there is:

1. proliferation of the uterine lining
2. increases in LH and FSH release and increased estrogen

End of the cycle includes:

1. inhibition of LH and FSH release
2. increase in progesterone
3. secretory phase in the lining of the uterus

Ovarian Cysts and Cancer

Cancers of the reproductive system are a constant threat to women. Because ovarian cancer can grow to a considerable size before any symptoms are noticed, it is particularly dangerous. About one in seventy women will develop this cancer, usually between the ages of fifty to seventy.

There are many cell types in the ovaries and so there are at least ten different types of ovarian cancer. It is also possible for ovarian cancer cells to spread directly to surrounding areas, through the bloodstream, or to other parts of the pelvis and abdomen.

Enlarged ovaries are a common sign of cancer, but this can also happen because of cysts and other conditions. As ovarian cancer develops, the abdomen may swell, a woman may feel pain, become anemic and lose weight.

An ultrasound, CAT scan, or viewing the ovaries with a laparoscope might help with diagnosis. Surgery is the immediate treatment, and radiation therapy and chemotherapy follow. If the cancer has spread to other organs, it is difficult to cure. However, early detection, successful surgery and the immune responses of different women can make a tremendous difference in survival rate.

Uterine Cancer

While cancer can occur in any part of the female reproductive system—the vulva, vagina, cervix, uterus, fallopian tubes or ovaries, cancer in the uterine lining is the most common cancer of the female reproductive system.

It usually develops after menopause and can spread, either down to the cervix or up the fallopian tubes to the ovaries, through the lymphatic system, or through the blood-stream. Abnormal bleeding is one of the early, common symptoms. One out of three women who have bleeding after menopause have this type of cancer, so this symptom requires prompt medical attention.

A hysterectomy, removal of the uterus, fallopian tubes, ovaries, and nearby lymph nodes, may be followed by chemotherapy. If the cancer has spread, progestins might prevent further growth.

- **If uterine cancer (endometrial carcinoma) is found early enough and treated, most women have a good prognosis.**

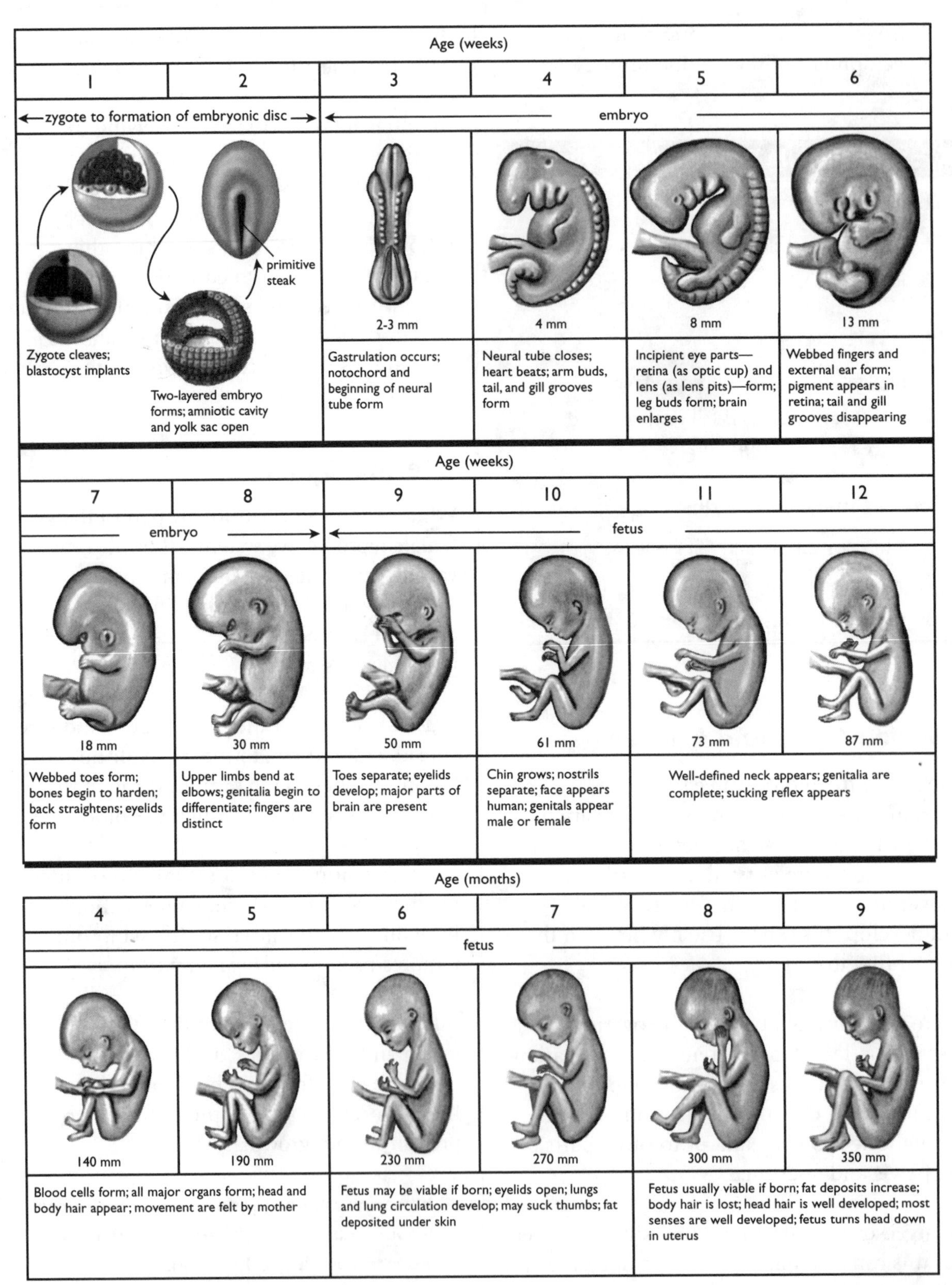

Figure 20.3—Human Embryonic Development

HUMAN DEVELOPMENT

When sexual intercourse happens at the time of a woman's ovulation, the conditions may be right for fertilization to occur. As a woman ovulates, her egg is released from the ovary and moves into the fallopian tube, where it might stay for up to three days.

After sexual intercourse, the sperm-laden semen swims up from the vagina, through the cervix and uterus and into the fallopian tubes. For fertilization to take place, a single sperm must break through the outer membrane of the ovum. The sperm's nucleus enters the egg's cytoplasm and fuses with the egg's nucleus. This new cell is called a *zygote* (see Figure 20.3).

- **Sperm may live for up to three days in the fallopian tube.**

The zygote moves through the fallopian tubes and progresses through a series of mitotic divisions called cleavage. It divides again and again and by the fourth day has become a solid ball of cells called a morula. It continues to divide, eventually becoming a hollow ball of cells called a *blastocyst*.

The blastocyst embeds itself into the uterine lining, and if this process is successful, it develops into a *gastrula* made up of three layers: the ectoderm, the mesoderm and the endoderm. The ectoderm becomes the nervous system, skin and sweat glands; the mesoderm becomes the reproductive system, kidneys, muscles, bones, heart, blood and blood vessels; and the endoderm forms the lungs, liver, linings of the digestive organs and some endocrine glands.

After gastrulation (transformation from blastocyst to gastrula), and the associated formation of three primary layers, the embryo has developed. It has two outer supporting membranes, the *amnion* and the *chorion*. The amnion provides a protective cushion around the embryo. The chorion, combined with some of the mother's endometrial cells, forms the placenta. This allows the developing fetus to exchange nutrients, oxygen and wastes with the mother. The embryo is connected to the placenta by the umbilical cord, which contains blood vessels going to and from the embryo.

- **Neurulation, the creation of the neural tube, gives rise to the central nervous system.**

After the ninth week of development, the embryo is called a *fetus*. At three months, it has started to develop many human features. The body systems are present, muscles move, the nervous system develops, blood cell formation starts and the fetus weights about as much as a greeting card.

At fourteen weeks, the fetus's hands, arms, legs and feet have reached birth proportions. At twenty-two weeks, the fetus is about thirty cm long and has regular sleep patterns. By thirty-two weeks, it is fully developed. The bones have hardened; the lungs and heart are ready for the outside environment. At about nine months, a surge in the pituitary hormone oxytocin begins the birth process. The uterus contracts more and more strongly, the amniotic sac breaks, and the cervix widens, allowing the infant to pass through the vagina.

SUMMARY

- The male reproductive system consists of the penis, scrotum and testes, and internally, the vasa deferentia, urethra, prostate gland and seminal vesicles. The testes produce the hormone testosterone, which contributes to secondary sex characteristic such as facial hair in males.
- Sperm cells develop in the testes, are stored in the epididymis, and are eventually ejaculated from the penis.
- Common disorders of the male reproductive system include prostate cancer, a slow-growing cancer that affects many men in their later years. Testicular cancer begins with a lump in the scrotum. It can be quite painful and removal of the one testis is usually the course of action.
- The female reproductive system is made up of the ovaries, the uterus, the vagina, urethra, and cervix. External genitalia include the vulva, the labia majora, and the labia minora.
- A narrow passageway in the cervix, usually impenetrable to sperm or other bacteria, opens during ovulation, allowing sperm to pass through to the fallopian tubes. From the ovaries, eggs are discharged into the fallopian tubes. If an egg meets with a sperm, and it is fertilized, then it can progress onto the uterus where it might become embedded in the endometrium, resulting in a pregnancy.
- Common disorders of the female reproductive system can include cancer of either the ovaries or uterus. Ovarian cancer is often hard to detect, and because there are so many kinds of cells in the ovaries, there can be many different types of cancer. Surgery, followed by radiation therapy and chemotherapy are the recommended treatment for ovarian cancer.
- Uterine cancer usually develops after menopause and it can easily spread down to the cervix, up to the ovaries, through the lymphatic system or through the bloodstream. Abnormal bleeding is one of the early, common symptoms. Treatment may include surgery and chemotherapy.
- When fertilization occurs, a single sperm breaks through the outer membrane of an ovum, creating a zygote. This zygote moves through the fallopian tubes and goes through mitotic divisions, called cleavage, and becomes a blastocyst embedded in the uterine lining.
- Gastrulations, the formation of three primary germ layers is followed by neurulation, the development of the central nervous system.
- The embryo begins to develop and at about nine months, it is fully viable in the outside world. The mother experiences a surge in oxtocin, a pituitary hormone, which begins the birth process. The uterus contracts more and more strongly, the cervix widens, and the infant emerges, head first, through the vagina.

CHAPTER 21

ECOLOGY

KEY TERMS

predation	deciduous	aestivation
decomposition	rain shadow	pheromones

Earth is a planet of extremes. Yet, no matter how hot or cold, wet or dry, life will gain a hold and even flourish. Millions of kinds of living things call this planet home. Ecology is the scientific study of the biosphere and its components. It is the study of the interactions of living organisms with each other and their environment. These interactions determine the distribution and abundance of organisms, where organisms are found, how many organisms occur there, and why the organisms occur where they do.

BIOSPHERE

The biosphere is the entire portion of the earth inhabited by life. The components of the biosphere are the hydrosphere, the lithosphere and the atmosphere.

- The **hydrosphere** is all of the water on earth. The water harbors the greatest quantity of life, with most of the organisms found in shallow waters along shorelines. The hydrosphere includes the salt water of the oceans, seas, and some lakes. It also includes the fresh water of rivers, streams, lakes, and ponds.
- The **lithosphere** is the rocky crust that forms the earth's rigid plates, and the terrestrial habitats. Most organisms live near the illuminated surface of the lithosphere.
- The **atmosphere** is gaseous, with nitrogen (seventy-eight percent) and oxygen (twenty percent) being the main constituents. Carbon dioxide makes up a tiny percentage of the air we breathe.

More heat and light strike at the equator because the light strikes perpendicular to the earth's surface. As you move north and south of the equator, there is a longer path for the sunlight because it is striking a curved earth. The biosphere is not uniform in width. The uneven distribution of solar radiation creates latitudinal differences in temperature and light intensity, and establishes vertical air currents resulting in three major air circulation areas on each side of the equator.

Ascending moist air near the equator releases moisture, and this is the area of the rainforests. Descending dry air about 30° north and south of the equator absorbs moisture, and this is the region of our great deserts. Cooling trade winds blow from east to west in the tropics and subtropics. Prevailing westerlies blow from west to east in the temperate zones. The permanent tilt of the earth on its axis causes seasonal variation in temperature and light intensity as the planet revolves around the sun (See Figure 21.1). The seasons are reversed in the north and south hemispheres.

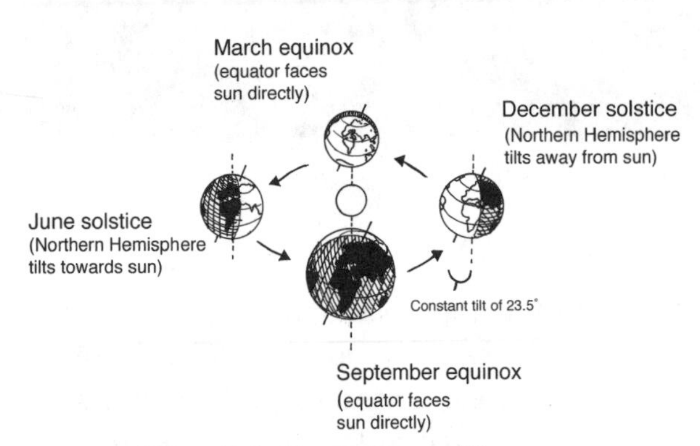

Figure 21.1—Earth's Tilts and Seasons

ECOSYSTEMS

Aquatic Ecosystems

Marine Biomes

Most of the surface of the earth is ocean. The open ocean is referred to as the *pelagic zone*. The pelagic zone is divided into three vertical layers:

- **The epipelagic zone** is the upper, sunlit, photosynthetic region of the ocean. Sunlight and the type of nutrients varies in this region, but nutrients are abundant. Large populations of phytoplankton flourish and provide energy and nutrients for other organisms.
- **The mesopelagic zone** does not get enough light for the process of photosynthesis. Large fish and other inhabitants of this region swim up to the epipelagic area to feed, or swim down to the ocean depths to feed on carcasses.
- **The bathypelagic zone** of the ocean is in continual darkness. No light-energy-absorbing organisms inhabit this zone. There are whole communities of organisms that inhabit the deep-sea thermal vents. Tube worms, sea anemones, sea stars, and other organisms inhabit this area of perpetual darkness.

The *neritic zone* lines the edges of continents or reef. The greatest concentration of marine life is located here. Enormous numbers of photosynthetic organisms inhabit this sunlit, nutrient-rich region of the ocean. Nutrients drain from the land into the ocean, and waves, tides, and winds distribute them. Upwelling (upward movement of water) provides fertile fishing grounds.

The *intertidal zones* are along the shorelines, and can either be rocky, where tide pools are formed, or non-rocky, where mud flats or sandy beaches form. In this region, there is a dynamic interplay between physical factors, such as temperature and dehydration. Competition and predation is high. The intertidal zone has four distinct strata: the splash zone, the high intertidal zone, the middle intertidal zone, and the low intertidal zone. Organisms that inhabit the uppermost stratum get the least amount of water. When the ocean is at high tide, these organisms are sprayed with the water. Those in the low intertidal zone are covered with water most of the time.

- **Less than three percent of the earth's water is fresh water.**

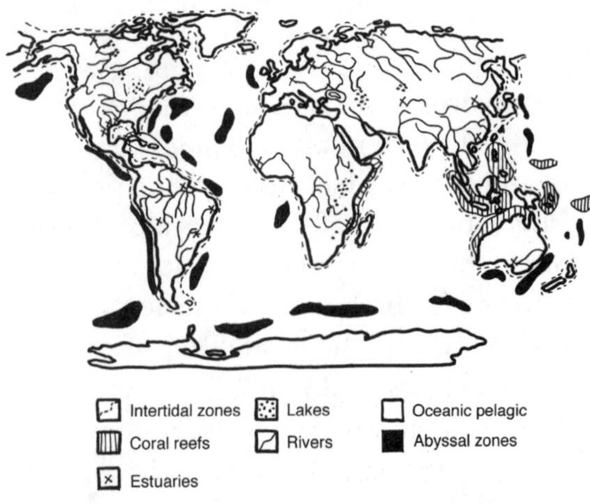

Figure 21.2—Earth's Major Aquatic Biomes

Rivers and Streams

Size determines the difference between a river and a stream. These bodies of water have a constant flow, which changes the shape of the land and causes unique habitats. The velocity of the flow affects the rate of erosion, the deposition of sediments, and the supply of oxygen, carbon dioxide, and nutrients. Friction along the bottom and the edges of the river affect the shaping and sculpting of the land over which the river flows.

Estuaries

An estuary is a unique environment where fresh water from rivers and streams empties into salt waters of an ocean. It is a mixing of salt and fresh water. The biggest physical factor in this zone is salinity. Estuaries are very fertile areas. Nutrients have been brought in from the rivers, and debris has been brought in by the tides. The nutrients become trapped due to tidal action and slow mixing.

Lakes and Ponds

Size determines the classification of lakes and ponds. They are standing bodies of water in depressions of the earth's crust. There are three different zones in a lake. The littoral zone is along the water's edge and the water is shallow. The limnetic zone is the lake's open, lighted water. The profundal zone is the deepest, darkest region of the lake.

Terrestrial Biomes

A biome is a large region of land characterized by habitat conditions and by its community structure. The predominant type of plants that grow there characterizes each biome.

Forests

Forests cover thirty percent of the earth's land surface. There are three types of forests: tropical rainforests, deciduous forests, and coniferous forests.

The tropical rainforests are found in a broad belt around the Earth's equator. They make up half of the earth's forests, and are found mainly in Central and South America, Africa, India, Asia, and Australia. Warm temperatures, constant day length all year long, and abundant rainfall characterize the rainforest.

- **The rainforest shelters more species of organisms than all of the other biomes combined.**

It is composed of an overstory with trees greater than fifty meters tall, and an understory of densely packed tall trees. Very little sunlight filters down to the forest floor because of the denseness of the trees. Where the light does reach the forest floor, ferns, shrubs, and mosses grow. The rate of decomposition is very fast in the rainforest. Dead organisms are attacked by hordes of bacteria, fungi, and nematodes (round worms). Nutrients become available for living organisms very quickly; there is no buildup of nutrients in the soil. Hence, the soil of the rainforest is not fertile.

Deciduous forests are composed mainly of trees that produce leaves during warm periods. Tropical deciduous trees lose their leaves during the dry season, and deciduous trees of the temperate zone lose their leaves during the cold season. There is an overstory of very tall trees, and an understory of trees whose growth and reproductive cycles coincide, with brief periods of maximum sunlight and favorable conditions. The deciduous forests are less dense than the rainforests and, therefore, sunlight reaches the floor. In drier regions, and those with low soil calcium, there is less variety of species. As you travel to the northern latitudes, and the weather gets colder, the number of species is limited even further.

Cold winters bring about a transition from deciduous forests to coniferous. Coniferous forests (inhabited by pines, fir, and spruce, for example) grow in colder climates and at the higher elevations. These forests have a dense overstory of trees. The number of different species among the conifers is minimal. The understory is composed of shrubs, ferns, and mosses. Due to the cold temperatures (the mean temperature is less than 0° C for six months or more), there is a short growing season and biological activity for about four months of the year. The forest floor is covered in pine needles, so the soil is acidic and infertile.

- **In the northern latitudes, the coniferous forests are called the taiga. Here the melting snows fill lakes. Watery bogs and marshes are found within many coniferous forests.**

Tundra

At the timberline, where trees thin out and eventually disappear from the landscape, you enter the coldest climates of the tundra. There are two types of tundra: the *alpine tundra* in the high elevations of mountain ranges, and the *arctic tundra* in high northern latitudes. Low-growing plants characterize them both. The plants include mosses, lichens, grasses, and dwarf shrubs.

The climate is brutal; during the warmest month of the year, the maximum temperature reaches about 10° C. In winter, there is thick snow and icy winds. In summer, there is some melting of the frigid marshes and ponds. The growing season is limited to about two to three months.

In the arctic tundra, only the top 0.5 meters thaws. The rest of the earth is permanently frozen soil (permafrost), which halts root growth, restricts drainage, and impairs decomposition. Only six hundred plant species are found in the entire arctic tundra.

Grasslands

Grasslands develop in regions with cold winters, hot summers, and seasonal rainfall. They are composed of densely packed grasses and herbaceous plants. As rainfall decreases from east to west, the height of the grasses diminishes and we observe tall grass prairies, mixed grass prairies, and short grass prairies. The soil is rich. This makes the grasslands a good choice for agriculture, grazing, and urbanization.

The savannas are grasslands with scattered or clumped trees. They have seasonal rainfall, punctuated by a dry season. During the dry season, above-ground stems are fuel for fast-moving fires. The fires usually do not affect the trees, and the rest of the savanna recovers quickly due to the growth of underground stems and roots.

Shrublands

Woody shrubs predominate in the shrubland biome. There is a Mediterranean type of climate, with hot, dry summers and cool, wet winters. The shrubs usually have small leathery leaves, few stomates, and thick cuticles. This combination of structures retards water vapor loss. The activity for most of the plants and animals in this region is restricted to the spring, when the temperatures are warm and the soil is still moist from the winter rains.

Deserts

Deserts comprise thirty percent of the land's surface. Intense solar heat, lashing winds, and little moisture characterize this biome. The great desert areas are located near 30° north and south of the equator due to the global air currents that create belts of descending dry air. Other deserts form in the rain shadows of tall mountain ranges.

There is only sparse vegetation. The plants of the desert have adaptations (tap roots and succulent leaves, for example) that allow them to survive the harsh conditions.

> Animals have adaptations for life in the desert. Some sleep in burrows during the day, and others sleep through the driest part of the year (aestivation). Animals that can perspire have a built-in mechanism for cooling off.

COMMUNITY ECOLOGY—INTERACTIONS BETWEEN ORGANISMS

Directly or indirectly, interactions among coexisting populations organize the communities to which they belong. Consider the following example. A species of New Guinea pigeons are fruit eaters. They digest the fruit and defecate the seeds. Where the seeds are dispersed will influence where new trees will grow. Tree distribution influences how the entire rainforest community is organized.

Competition is an interaction that harms both organisms. When a shared resource is plentiful, there is enough for all. Most resources are limited, however, so organisms with overlapping needs cause increased competition. When members of the same species compete for a resource (intraspecific competition), the competition can be more intense than between members of different species (interspecific competition).

During indirect competition, the competitors have equal access to the resource, but the faster or stronger species obtains more of it, reducing the competitor's supplies. For example, tamarisk trees planted as windbreakers in the California desert are killing the native mesquite and desert willow trees. The tamarisks are more efficient at tapping water, and they grow more rapidly than their competitors.

Direct competition among organisms is demonstrated by aggressive animal behavior (hyenas drive away vultures from dead animals, for example), and territoriality of animals (male bighorn sheep defend their group from other male bighorn sheep).

The winner takes all. One of the species has a selective advantage (adaptation) which allows it to be more successful than the other. This adaptation increases the chances of survival and, therefore, reproduction of the organism. Daylilies afford a good example of this phenomenon. They form thick clumps of shoots and roots. Few native plants can compete with daylilies, which take over any new community by crowding out the native species.

> In 1957, goats were brought onto Abingdon Island, in the Galapagos Islands. The goats ate low-growing plants and the island's native tortoises lost their food supply. When the researchers came back in 1962 to monitor the goats, the native tortoises had become extinct!

Is there an alternative to to competition? Yes, if competitors can use different parts of the same resource. This is called *resource partitioning,* and can occur when animals inhabit different parts of the same tree or pond (*spatial partitioning*) or when two plants grow during different seasons (*temporal partitioning*).

Intense competition may lead to changes in one or more characteristics of a species. If birds of two species, for example, ate the same size and kind of fruit, and one of the species evolved into a group that could eat larger fruit, competition would be reduced. This is referred to as *character displacement*.

Interactions that harm one organism and benefit the other include *predation* and *parasitism*. All organisms need nourishment. *Predation* occurs when one organism (predator) acquires its needed resources by eating another organism (prey). Predators may be herbivores, carnivores, or omnivores. Coevolution occurs between the predators and their prey. Increased skill of the predators in capturing their prey is often balanced by improved methods of escape on the part of the prey.

Many organisms have escape adaptations. In some instances, the prey avoids detection due to camouflage, making them difficult to detect. In cryptic coloration, the color of the prey looks like the background. For example, some hares are white in winter (difficult to see against a snowy background) and become brown in the summer.

In addition to coloration and shape, behavior can contribute to a successful disguise. Some animals like the opossum, remain motionless when faced with a predator. Owls can fluff up their feathers and startle the predator.

Some lizards can detach their tails and escape capture. Other prey can outrun the predator. Group responses can warn or protect the group, and confuse the predators. Schools, packs, herds, and colonies of prey gain their defense in numbers. The herd, typically, surrounds the weak of their species.

Physical defenses protect a variety of prey against their predators. Mollusks and turtles have shells. Porcupines are equipped with quills. Many plants have piercing thorns, spines, or stinging hairs, which can deter a predator. Organisms can be foul–tasting, poisonous, stinging, and smelly, or have a nasty bite.

- **Certain foul-tasting moths make a clicking sound and bats learn to eat quiet moths!**

Some organisms protect themselves with chemical defenses, such as noxious propellants, or painful or deadly bites and stings. The larva of the Monarch butterfly feed on milkweed, which has toxic chemicals in it. The chemical is passed on to the butterfly during metamorphosis, which makes the Monarch butterfly poisonous. The skunk's chemical smell discourages predators.

In *parasitism*, one organism is harmed, while the other benefits by the interaction. A look at the human tapeworm will serve as an example of this phenomenon. If we eat undercooked pork or beef that is infested with the tapeworm, the organism will enter our digestive tract. This worm lacks eyes, a digestive tract, and a muscle system. Its outer coat protects it from the acid of the human stomach. Its long, flat shape gives it a maximum absorbing area, but does not obstruct the host's intestine, its niche. The "head" is called a scolex and contains hooks to allow it to attach and anchor itself to the intestinal wall. The parasite is nourished by the host, which usually lives but may feel ill.

An example of social parasitism is seen with the European cuckoo. A host bird builds a nest and lays its eggs. The European cuckoo goes to the nest and destroys one host egg, replacing it with one of its own. The host bird sits on the eggs until they hatch. The cuckoo egg hatches first, and the hatchling throws out all solid objects (including the host eggs) from the nest. It then eats the food brought in by the foster parents!

In *mutualism*, both benefit, and interaction is necessary for survival or reproduction of both organisms. The pollination of some flowers by specific insects, birds, or bats displays this type of interaction. The relationship between termites and protozoa is mutualistic. It is the protozoa in the termite's gut that digest the cellulose of the wood that the termites ingest. The termites get useable nutrients, and the protozoa are given a habitat and food supply. One would not survive without the other.

Commensalism is a relationship in which one organism is benefited, and the other is unaffected. There are many examples of this interaction in nature. We see barnacles encrusted on the humpback whale. They get a habitat and transportation to a new food source. The whale is unaffected.

A close, long-term relationship between individuals of two different species is called *symbiosis*. Fungi and algae may form a symbiotic relationship. The algae are sandwiched between two outer layers of fungal filaments. Together, as lichen, they thrive in a wide variety of habitats and conditions, where, sometimes, neither could grow alone. In dry climates or in places poor in organic nutrients, the photosynthetic algae uses the light-energy to generate organic compounds from inorganic compounds. The fungal filaments gather and conserve what little water is available. Their combination is perfect in harsh, cold climates, where lichens can support the food chain. In the arctic, lichens sustain large consumers, such as reindeer caribou.

A bat can actually link two ecosystems together, for example, a lake and a cave. As a lake fills with sediment, more plants grow. Insects thrive as more plants grow around the lake. The bats eat insects around the lake. After eating, the bats return to the cave and defecate. The fecal material becomes food for fungi in the cave. The fungus, in turn, is a nutrient for cave beetles and other insects. Thus, the bat links a lake ecosystem with a cave ecosystem.

POPULATION ECOLOGY

A population is a group of individuals of the same species occupying a given area. They rely on the same resources, are influenced by similar environmental factors, and have a high likelihood of interacting with each other. The demographics, or vital statistics, of a population include population size, density, distribution, and age structure.

The population size is the number of individuals that contribute to a population's gene pool. Increases in population are due to births and immigration of new individuals to an area. Decreases in the population size occur due to deaths and emigration of individuals.

The number of individuals per unit area or per volume is the density of the population. How do you count individuals? In rare instances, you can count all the individuals of a population directly. The human census is a direct count. More frequently, the number of individuals is counted by using sampling techniques to estimate the total population size.

There are three patterns of population dispersion: clumped, uniform, and random spacing. The *clumped* pattern of dispersions is the most common. Plants may be clumped in certain sites where soil conditions and other factors favor germination and growth. Some animals, such as insects and salamanders, clump together under logs where the humidity is high. Clumping of animals is associated with social behavior, such as the swarming of insects, flocking of birds, and schooling of fish.

In a *uniform* distribution, the individuals are more evenly spaced. For example, a tendency toward regular spacing of plants may result from shading and competition for water and minerals. Some plants secrete chemicals into the soil, which prevent other individuals of the same species from germinating, reducing the competition for water and minerals. This occurs with the oak tree. Uniform dispersion may be the result of social interactions, such as birds nesting on an island as each bird needs a certain amount of space.

The age and sex of individuals affects the population growth. Population biologists plot the number of individuals of a certain age and sex to determine the age-sex structure for a population. When a large proportion in a population are at a reproductive age or younger, the age-sex structure tends to be shaped like a pyramid. Developing countries have a pyramid-shaped age-sex structure, and their populations are increasing dramatically (See Figure 21.3).

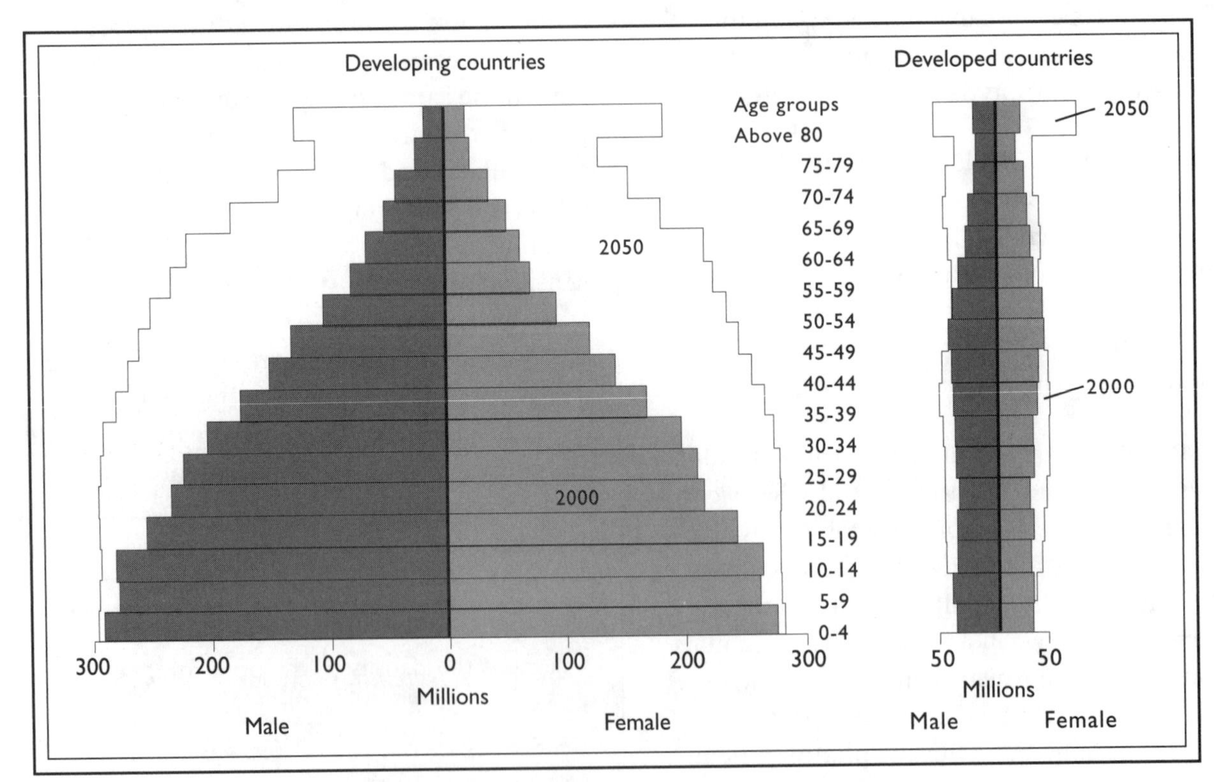

Figure 21.3—Comparing Projected Population
2000 population profiles for developed and developing countries projected to the year 2050 (Source: U.S. Census Bureau, International Data Base)

There is less growth of the population when the number of individuals capable of reproduction is less than the older population.

The maximum rate of increase per individual under ideal conditions is referred to as the *biotic potential*. There is a maximum rate for each species. No organism grows faster than bacteria. For bacteria, the biotic potential is one hundred percent every twenty minutes!

- **If the ideal rate of growth for bacteria could continue for thirty-six hours, there would be enough bacteria to form a layer one foot deep over our entire planet.**

No natural ecosystem can support continuous exponential growth for any species. No ecosystem has unlimited resources, and environmental conditions never remain constantly favorable for limitless growth. The carrying capacity is the actual number of individuals in a population that a given environment sustains indefinitely. Populations typically start out with a low density, and grow slowly. For a time the growth is very rapid, and then it levels off in size once the carrying capacity has been reached.

What are the limiting factors on the growth of populations? A limiting factor is any essential resource that is in short supply, which will limit the population. If you sustain the supply of necessary resources, this will be a major factor in determining population size.

Environmental resistance to population growth is caused by one or more factors that eventually limit the population growth. These include competition, predation, severe weather, disease, limited food and/or water, restricted space, depleted soil nutrients, or toxic by-products of the organisms themselves.

ANIMAL BEHAVIOR

Ethology is the study of animal behavior. Genetics plays an important role in determining the actions of animals that have little opportunity to learn. Behaviors that are precisely specified by genes are often those that must be nearly perfect the first time (e.g., the ability to escape from a predator). Behavior is adaptive, that is, it enhances the animal's chances of surviving and reproducing.

Primarily *innate behavior* is under precise genetic control, often species-specific and highly stereotyped. If a salamander is raised away from water until long after its siblings begin swimming successfully, it will swim as well as they the first time it is placed in the water. Another example of innate behavior is the release of pheromones, chemical signals released by many animals. These can elicit alarm behavior, mating behavior, and foraging behavior in other members of their species.

Animals may "learn from experience." Habituation is the simplest form of learning. Let the scarecrow in a corn field serve as an example. When the scarecrow is first placed in the field, birds will fly away. In a short period of time the birds will recognize that it is not a threat, and will even perch on the scarecrow!

Insight learning is a sudden solution to a problem without obvious trial and error. Is it the result of associations among previously learned components? Consider this example. Chimpanzees are in a cage. A bunch of bananas is hanging from the ceiling, too high for the chimps to reach. There are empty boxes in the corner of the cage. The chimps stack the boxes under the bananas, and climb up to reach and eat them.

Animals learn from others in *social learning*. Is it more efficient to learn from others? Rhesus monkeys learn to fear and avoid snakes by watching the behavior of other monkeys. Juvenile wrens need to countersing with an adult wren to learn the song for territorial defense. Examples of social learning among humans are limitless.

Imprinting occurs during a restricted period of time, without reinforcement. This is demonstrated with chicks that follow their mother wherever she goes. These birds follow the first thing that moves after they are born, and, that is usually their mother.

Communication among animals is essential to social life. Animals may use visual signals, such as color or brightness during the day for short-range communication and in the open environment. Sound communication is effective during the day or night, and is usually employed where vision is limited (in dense foliage, for example). Sound communication is very effective in water, as well as on land. Many social animals communicate via the sense of smell with pheromones. These chemicals are used to communicate over long distances, and can mark the territories of animals.

Social organisms live in groups. What are some of the benefits of group living? Groups of animals can detect, confuse, and repel predators. Many exhibit a mobbing behavior with their predators. Groups of small birds can overtake an owl. Are there costs to the animals for their group living? There is increased competition for mates, nest sites, and food. There is increased exposure to parasites and disease. They are more conspicuous to their predators or prey. As always in nature, there seems to be a plus and a minus to every adaptation.

SUMMARY

- The hydrosphere, the lithosphere, and the atmosphere support all life on earth, The greatest quantity of life is found in the oceans. Most of the life in the oceans is found in the shallow coastal waters.
- The biomes of the land are characterized by the major plants that inhabit them. A canopy of tall trees characterizes the forest biomes. Other biomes include savannas and shrublands. Deserts have the least amount of vegetation. The tundra regions have low-growing plants and shrubs adapted to this hostile region.
- Organisms interact with members of different populations. Competitions among organisms exist at various levels. If the competition is extreme, one species may become extinct. Predation and parasitism are interactions that harm one organism and benefit the other.
- Population ecology is the study of the number and density of individuals in an area. The number of individuals is a direct relation to the number of individuals capable of reproducing. Density is affected by environmental conditions.
- Animal behaviors can range from completely innate to those that are learned. Learning is a process by which the animal benefits from its experience.

CHAPTER 22

THE FUTURE OF BIOLOGY

KEY TERMS

laparoscopy	microsurgery
twinning	chondrocyte

At the turn of the twentieth century, there were no such procedures as blood transfusion or organ transplant. There were no antibiotics. The structure of DNA was a mystery. Just after World War I, an influenza epidemic claimed more victims than the war did. In the 1950s, before Jonas Salk developed a preventive vaccine, thousands of people died or were permanently disabled by polio. Consider how far the science of biology has progressed in the last one hundred years. What wonders are being discovered in the laboratories of the twenty-first century? Where will science take us in this millenium?

BIOLOGY TODAY

The medicine that was practiced twenty years ago would seem primitive by today's standards. Many treatments were just a "shot in the dark," and operating room procedures did not compare with modern laparoscopy and microsurgery. It was less than twenty years ago that the U.S Food and Drug Administration approved the first genetically engineered drugs, such as Herceptin (used to treat breast cancer).

THE ETHICS OF PROGRESS

It is true that the first fifty years of the twentieth century saw some amazing discoveries and great progress in human biology, but the second half of the century has brought us all the way from the polio vaccine to the Human Genome Project. Humans have cracked the DNA code. Scientists have already moved on from the realization of how DNA provides the instructions for a human being, to proteomics, the study of all proteins whose synthesis is coded for by genes.

Much of the research of the last half of the twentieth century focused on nucleic acids, large complex molecules that contain hereditary, or genetic, information. Once the double-helix structure of DNA was discovered, the science of genetics moved swiftly forward. Today, we face a multitude of dilemmas that simultaneously involve issues related to ethics, politics, economics, and medicine. It is one thing to be able to use genetic screening to help prevent a terrible birth defect; it is entirely another thing to select a child based on eye color or intelligence. Controversy awaits at every turn, as scientists weigh when, and how, to use gene therapy to cure or prevent illness, to predict someone's future, and even to create new life.

What comes next is a tough question for researchers, physicians, and the general public.

When is it right to use genetic engineering? When, perhaps in terms of germ warfare, does it become unethical? Who is to decide?

Along with great strides in medicine has come the ability to terrorize whole populations with the threat of chemical or biological agents. As our knowledge of biology has increased, so has the ability to create weapons of mass destruction. The use of Agent Orange in Vietnam and other chemical agents in the Gulf War, has already set the stage for the future. Aerosol sprays can disperse biological agents, if inhaled, but they can also be used to contaminate food or water. Some chemical agents are volatile, evaporating rapidly to form clouds. Others may act directly on skin, lungs, eyes, or the respiratory tract. The terrorist threats and confirmed cases of exposure to anthrax, in late 2001 and early 2002, gave way to even more widespread anxiety about the possibility of a terrorist attack using smallpox, plague, or other chemical or biological agents.

While the world debates the ethical issues surrounding genetics and germ warfare, modern medicine has made incredible advances in prevention and treatment of many diseases, in surgical procedures, new technology, and drug therapies.

CLONING

A fascinating topic in modern biology is that of cloning. Dolly, the first cloned sheep, was created at the Roslin Institute in Scotland in 1997. Dolly was created by transferring genetic material from the nucleus of an adult sheep's udder cell to an egg whose nucleus, and thus its genetic material, had been removed (See Figure 22.1). Dolly, therefore, carried the

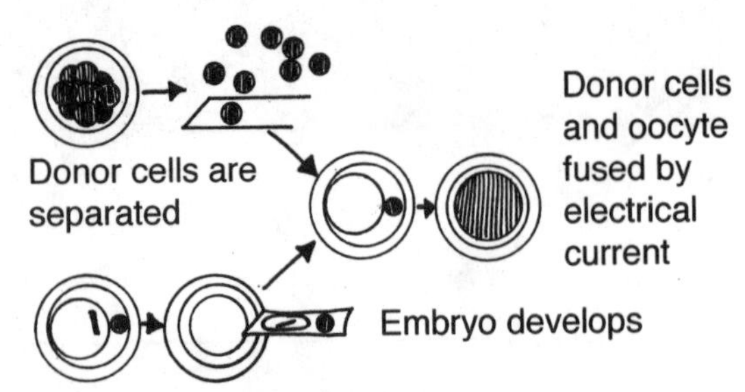

Figure 22.1—Cloning Procedure

DNA of only one parent. Cloning also refers to copying genes and other pieces of chromosomes to generate enough material to study, and to blastomere separation, or *twinning*, when a developing embryo is split soon after the fertilization of the egg.

Dolly preceded a number of other clonings of small animals, and heralded a heated controversy about the ethics of cloning. The ability to clone, however, gives scientists an unprecedented opportunity to study the possibilities of genetically altering animals, and reproducing them with predictable results. Researchers explore the ramifications that cloning might have on human disease, and consider raising animals with genetically altered organs suitable for human transplants.

Dolly, suffering from lung disease and arthritis, was put to sleep in 2003, about four years short of a normal sheep's life span. One of the questions that remains about cloning is how the genetic changes in cells used to obtain nuclei might affect the cloned animal's health.

NEW DEVELOPMENTS

Vaccines against the flu, smallpox, polio, measles, and mumps have changed our lives over the past century. Today, new vaccines that can help turn on the body's immune system in the fight against cancer offer continued hope for the future.

Research shows that modified dendrite cells, particularly from Prostate-Specific Antigen (PSA) and Carcinoembryone Antigen (CEA) proteins, can activate the immune system. It is now possible to use a tumor's genetic RNA to identify more cancerous cells than is possible with a single-antigen vaccine. It is likely that RNA-based vaccines will have the potential to help people who would otherwise have no options other than surgery, chemotherapy, and radiation.

Genetic engineering of drugs is another area that looks promising for the future. Drugs are being developed against diseases like malaria, tuberculosis, and HIV/AIDS, with the potential to save millions of lives. The issue is to produce such drugs in a cost-efficient manner so that many of the world's poorer populations will have access to them as well.

Genetic Engineering

The understanding of genetics has also made genetic counseling available to people in a variety of ways. It might involve the accurate diagnosis of a disorder, help with understanding the risk of recurrence in a family, and help with decision-making when it comes to reproduction. Research in genetics has given us an unbelievable number of options in determining the sex and characteristics of a child. Still, although there is the possibility of selecting genes or replacing genes that cause disease, these options are not always affordable, or even practical. Genetic counseling is more about knowing what your risks are and making life choices that are appropriate.

- **There are approximately 2,000 genes identified, and about 5,000 disorders caused by genetic defects.**

Counseling can be particularly important in the field of obstetrics. To date, there are no cures for Down syndrome or Huntington's disease, but genetic counseling might alert a family to these possibilities. A complete family history, genetic tests using DNA analysis, X-ray, ultrasound, urine analysis, skin biopsy and physical evaluation can help people with complex family histories, or prospective parents, make educated decisions.

Implants

In the 1970's, metal implants, with new metals and structural implants for joint replacement, dominated orthopedics. Since that time, orthopedic researchers and physicians have discovered therapies that stimulate bone marrow to form repair tissue. In one technique, autologous chondrocyte transplantation, a surgeon obtains a biopsy of healthy knee cartilage. This is delivered to a company that harvests approximately twelve million cells. About a month later, the surgeon removes damaged tissue and inserts cultured cells, using tissue from the lower leg to suture over the surgical site. The cultured cells multiply, integrating with surrounding cartilage to produce hyaline-like cartilage.

The success rate for autologous chondrocyte transplantation, according to researchers, is high, and as many as eighty-five percent of patients exhibit good to excellent results. Many were difficult cases to begin with, with advanced disease or degeneration of the bone.

High Resolution Imaging

Another life-saver in modern medicine is high resolution imaging. For instance, a High Resolution CT of the lung (HRCT) has given physicians a great advantage over the typical chest radiography and clinical studies. The HRCT uses a very narrow X-ray beam and a "high-spatial frequency reconstruction algorithm" to provide extremely high definition images of the lung, the vessels, airspaces, and airway. The high resolution images allow more accurate diagnosis of a variety of lung diseases, and thus allow physicians to order the most appropriate treatment.

Artificial Intelligence

Even harder to imagine, perhaps, is the study of artificial intelligence. If intelligence is defined as the computational aspect of the ability to achieve goals, then the creation of computers that are intelligent is, of course, plausible. The ultimate goal is to create machines with human-level intelligence, but because fundamental new ideas may be needed, rather than more complex programs, it is hard to predict when this could happen. The idea of a non-biological machine with intelligence is probably no more remarkable than the miracle of cloning. It is sometimes the speed with which such advances are made that leave the non-scientific public gasping in awe.

THE FUTURE

What the future holds for biology is beyond our imagining. Now that animals have been cloned, what is next? If genetically engineered drugs and cancer vaccines can help to eliminate some of the world's serious diseases, and if joints and organs can be transplanted and replaced with greater ease, how long might people actually live?

Germ-line genetic engineering, repairing individual genomes by replacing defective genes, might give us tremendous power; it might also give us the ability to prevent or cure otherwise intractable diseases. We will be able to change the course of our lives, and the lives of our families, to cure diseases that once seemed impossible to overcome. But, at the same time, genetic engineering promises to open new mysteries, introduce new horrors, to lead us in directions that we can't even begin to fathom at this point.

The Human Genome Project was kicked off in 1990, finished in 2000. Embryos were cloned in 1993. Dolly the sheep was produced in 1997, a house cat in 2002. In 2003 we celebrate the 50th anniversary of the discovery of the double helix of DNA. Imagine what will have happened in another ten, twenty, or thirty years, and the impact that such discoveries might have on our lives.

APPENDIX I

THE GENETIC CODE

First Position (5'-end)	Second Position: U		Second Position: C		Second Position: A		Second Position: G		Third Position (3'-end)
U	UUU	Phe	UCU	Ser	UAU	Tyr	UGU	Cys	U
U	UUC		UCC		UAC		UGC		C
U	UUA	Leu	UCA		UAA	Stop	UGA	Stop	A
U	UUG		UCG		UAG		UGG	Trp	G
C	CUU	Leu	CCU	Pro	CAU	His	CGU	Arg	U
C	CUC		CCC		CAC		CGC		C
C	CUA		CCA		CAA	Gln	CGA		A
C	CUG	C	CG		CAG		CGG		G
A	AUU	Ile	ACU	Thr	AAU	Asn	AGU	Ser	U
A	AUC		ACC		AAC		AGC		C
A	AUA		ACA		AAA	Lys	AGA	Arg	A
A	AUG	Met or Start	ACG		AAG		AGG		G
G	GUU	Val	GCU	Ala	GAU	Asp	GGU	Gly	U
G	GUC		GCC		GAC		GGC		C
G	GUA		GCA		GAA	Glu	GGA		A
G	GUG		GCG		GAG		GGG		G

PERIODIC TABLE OF THE ELEMENTS

Period	Group	21	3	4	5	6	7	8	9	10	11	12	13	14	15	16	17	18
1	1 **H** 1.008																	2 **He** 4.003
2	3 **Li** 6.941	4 **Be** 9.012								Atomic Number **Symbol** Atomic Mass			5 **B** 10.811	6 **C** 12.010	7 **N** 14.007	8 **O** 15.999	9 **F** 18.998	10 **Ne** 20.178
3	11 **Na** 22.990	12 **Mg** 24.3050											13 **Al** 26.982	14 **Si** 28.0855	15 **P** 30.974	16 **S** 32.066	17 **Cl** 35.453	18 **Ar** 39.948
4	19 **K** 39.098	20 **Ca** 40.078	21 **Sc** 44.9559	22 **Ti** 47.867	23 **V** 50.942	24 **Cr** 51.9961	25 **Mn** 54.938	26 **Fe** 55.845	27 **Co** 58.933	28 **Ni** 58.6934	29 **Cu** 63.546	30 **Zn** 65.39	31 **Ga** 69.723	32 **Ge** 72	33 **As** 74.922	34 **Se** 78.96	35 **Br** 79.904	36 **Kr** 83.80
5	37 **Rb** 85.468	38 **Sr** 87.62	39 **Y** 88.906	40 **Zr** 91.224	41 **Nb** 92.906	42 **Mo** 95.94	43 **Tc** [98]	44 **Ru** 101.07	45 **Rh** 102.906	46 **Pd** 106.42	47 **Ag** 107.868	48 **Cd** 112.411	49 **In** 114.818	50 **Sn** 118.710	51 **Sb** 121.760	52 **Te** 127.60	53 **I** 126.904	54 **Xe** 131.29
6	55 **Cs** 132.906	56 **Ba** 137.327	57-71 Lantha-nide Series	72 **Hf** 178.49	73 **Ta** 180.948	74 **W** 183.84	75 **Re** 186.207	76 **Os** 190.23	77 **Ir** 192.217	78 **Pt** 195.078	79 **Au** 196.967	80 **Hg** 200.59	81 **Tl** 204.383	82 **Pb** 207.2	83 **Bi** 208.980	84 **Po** [209]	85 **At** [210]	86 **Rn** [222]
7	87 **Fr** [223]	88 **Ra** [226]	89-103 Actinide Series	104 **Rf** [261]	105 **Db** [262]	106 **Sg** [263]	107 **Bh** [262]	108 **Hs** [265]	109 **Mt** [266]	110 **Uun** [269]	111 **Uuu** [272]	112 **Uub** [277]	113 **Uut** [Undiscovered]	114 **Uuq** [285]	115 **Uup** [Undiscovered]	116 **Uuh** [289]	117 **Uus** [Undiscovered]	118 **Uuo** [293]

Series															
Lantha-nide Series 57-71	57 **La** 138.906	58 **Ce** 140.116	59 **Pr** 140.908	60 **Nd** 144.24	61 **Pm** [145]	62 **Sm** 150.36	63 **Eu** 151.964	64 **Gd** 157.25	65 **Tb** 158.926	66 **Dy** 162.50	67 **Ho** 164.930	68 **Er** 167.26	69 **Tm** 168.934	70 **Yb** 173.04	71 **Lu** 174.967
Actinide Series 89-103	89 **Ac** [227]	90 **Th** 232.038	91 **Pa** 231.036	92 **U** 238.029	93 **Np** [237]	94 **Pu** [244]	95 **Am** [243]	96 **Cm** [247]	97 **Bk** [247]	98 **Cf** [251]	99 **Es** [252]	100 **Fm** [257]	101 **Md** [258]	102 **No** [259]	103 **Lr** [262]

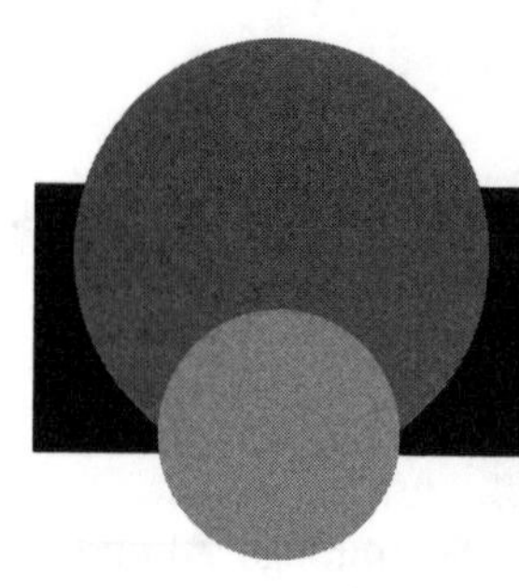

GLOSSARY

actin
a protein that acts as contractile material in muscle fiber.

adaptation
a beneficial change in an organism that allows it to survive in new conditions.

aestivation
a state of inactivity in an animal during which bodily functions, such as respiration, slow.

allele
one form of a gene. In a gene there are usually two alleles in the same relative position on homologous chromosomes.

allergen
an antigen that induces an immune response.

anabolic
relating to the building of complex chemical compounds from simpler compounds.

anticoagulant
a substance that prevents blood clotting.

antigen
a substance (often a protein) that induces an immune response.

atom
the smallest form of an element that still has the traits of that element.

autonomic
the part of the nervous system which regulates movement of smooth muscle, cardiac muscle, and gland cells.

basophil
a phagocytic white blood cell.

binary fission
simple cell division in which the two new cells are alike and equal in size.

biodiversity
the existence of a wide variety of organisms in a given environment.

biology
the study of living organisms and life processes.

blastocyst
a hollow ball of cells of the blastula stage in the development of a mammalian embryo.

bolus
a chewed ball of food ready to be swallowed.

bronchitis
inflammation of the mucous membrane of the bronchial tubes.

camouflage
the ability of an animal to visually blend with its surroundings.

capsid
the protective protein coat of a virus

capsomere
a part of the protein coat of a virus.

carbohydrate
organic compound containing carbon, hydrogen, and oxygen. Examples of carbohydrates are simple sugars, starch, glycogen, and cellulose.

catabolic
relating to the breaking down in the body of large chemical compounds into smaller ones.

catalyst
a substance that accelerates a chemical reaction but is not used up or permanently changed in the process.

cell
the smallest functional independent unit of a living organism.

chemical bonds
an attractive force holding atoms together.

chondrocyte
a nondividing cell of the matrix of cartilage.

chromatid
each of two threadlike strands formed by a chromosome during mitosis or meiosis. Each chromatid becomes a daughter chromosome.

chromosome
a structure in the cell nucleus that carries the genes and is capable of reproduction through cell division.

codon
three consecutive nucleotides in a strand of DNA or RNA that genetically codes for a specific amino acid.

coenzyme
a substance that facilitates the action of enzymes.

compound
a substance formed by the bonding of two or more elements.

corpus luteum
a structure formed in the ovary at the site of a burst ovarian follicle.

cytokine
protein substance secreted by certain white blood cells, which helps to regulate immune response and acts as a signal in cell-to-cell communication. Examples are interferon and interleukin.

cytokinesis
change in the protoplasm of a cell during cell division.

cytoplasm
the material within a cell that includes the gel-like fluid and organelles, but not the nucleus.

deciduous
the term used to describe plants that lose their leaves at the end of the growing season.

decomposition
the chemical breakdown of matter to a simpler state.

deoxyribonucleic acid
the genetic material within a cell that is also known as DNA.

diffusion
uniform distribution of molecules throughout a substance due to their random movement.

diploid
a cell containing two sets of chromosomes, one from the father and one from the mother.

distal
situated away from the center of the body.

dominant
in genetics, describing the allele that is usually expressed when two alleles are present.

duct
a tubular structure through which a substance may be secreted or excreted.

ectotherm
an animal with an externally regulated body temperature.

element
a substance composed of one kind of atoms.

endocytosis
a process by which a material enters a cell without passing through the cell membrane. The membrane folds around the material creating a small sac that pinches off, isolating the material within the cell.

endometrium
the mucous membrane that makes up the inner layer of the uterus.

endotherm
an animal with an internally regulated body temperature.

environment
the external conditions, physical, chemical, and biological, in which an organism exists.

enzyme
a protein catalyst, which causes chemical change in other substances, without being changed by the process.

epithelium
a cellular layer that covers the outer surface of the body, walls of internal cavities, and the glands.

ethmoidal
relating to the air cells with thin bony walls that are found in the nasal cavity.

exocrine
describing a gland that secretes a substance into a body cavity or onto a body surface through a duct.

exocytosis
the process by which substances leave a cell through the cell membrane in vesicles and are then released to the exterior.

fibroblast
a spindle-shaped cell in connective tissue.

follicle
a spherical cluster of cells, usually with a cavity within that may hold a cell or other structure.

gamete
a reproductive cell.

ganglia
a collection of nerve cells usually in the peripheral nervous system.

gastrulation
the change of the blastula into the gastrula.

genome
a complete set of chromosomes acquired from one parent.

genotype
the genetic makeup of an organism.

glycerol
a sweet, oily, trihydric alcohol.

haploid
describing a reproductive cell with a single set of unpaired chromosomes.

hemoglobin
the protein substance of mature red blood cells that transports oxygen from the lungs to the tissues.

heterotrophic
describing a different type or form.

heterozygous
allelic genes at one locus that are different.

histamine
a vasodilator, generally released during an allergic reaction, that constricts bronchial smooth muscle, increases gastric secretion, and may precipitate a fall in blood pressure.

histone
water-soluble protein that contains a large proportion of basic amino acids.

homologous
describes chromosomes with the same structural features.

homozygous
allelic genes at one locus that are the same.

hyphae
filaments of fungi which may form a loose network or a tightly woven mass.

intercalated disk
a specialized structure that connects two adjacent cells.

interferon
a class of glycoproteins released by a virus-infected cell, that communicates with neighboring cells and disrupts viral translation.

interstitial
extracellular space.

keratinocyte
a cell that produces keratin, the material that generally makes up hair, nails, and horns.

kinin
polypeptide signalling molecules formed in blood that influence smooth muscle contractility.

lacteal
a lymphatic vessel that takes up chyle from the intestine and conveys it to the lymphatic system.

laparoscopy
examination of the contents of an internal cavity such as the abdomen, with a laparoscope.

lipid
a compound that is not soluble in water, but is soluble in organic solvents. There are two main categories of lipids: complex lipids, or fatty acids (such as glycerides and phospholipids), and simple lipids, such as terpenes and steroids.

locus
the position of a gene on a chromosome.

macrophage
a phagocytic white blood cell developed from a monocyte.

mast cell
a connective tissue cell, which aids the arrival of leukocytes at a site of infection; secretes histamine, and is important in allergic responses.

matrix
the surrounding substance within which a particle or structure is embedded.

matter
everything physical that makes up the universe.

mediastinum
the median partition of the thoracic cavity.

melanin
any of a group of polymers derived from tyrosine that cause pigmentation in skin, hair, and eyes.

membrane
a layer of tissue covering, lining, or connecting structures.

metabolic rate
a measure of the energy used by an animal within a specific period of time.

microfilaments
the thinnest filaments of the cytoskeleton.

microsurgery
surgery performed with a surgical microscope.

mucous
a substance secreted by goblet cells to lubricate a membrane and trap foreign bacteria.

mutation
a mistake during the copying of DNA in which the sequence of a gene is altered, creating new alleles which may enter the gene pool.

myelin sheath
an insulating outer layer in vertebrates surrounding axons.

myosin
a protein that acts as contractile material in muscle fiber. It forms the thick filaments in muscle.

neurotransmitter
a signalling molecule released by a nerve cell or gland, that stimulates other cells.

niche
in ecology, the position occupied by a species in its community.

nucleic acid
a family of macromolecules in all cells, and viruses.

organelle
a specialized structure in a cell that carries out certain functions, the endoplasmic reticulum, for example.

organic
relating to a carbon compound other than an oxide or carbonate.

organism
any individual living thing.

pathogen
a bacterium, virus, fungus or parasite that causes disease.

peptide
a compound of two or more amino acids linked by peptide bonds.

perfusion
the flow of a fluid, such as blood, per unit volume of tissue.

pepdidoglycan
a compound containing amino acids, that strengthens the cell wall of bacteria.

pericardium
the membrane that encloses and holds the heart in position.

peristalsis
waves of contraction and relaxation along a tubular structure, such as the intestine.

phagocyte
a white blood cell that ingests pathogens and other cells.

phenotype
the expressed traits of a genotype.

pheromone
chemical message secreted by an individual and perceived by another individual of the same species, often altering the behavior of the second individual socially or sexually.

phloem
tissue in vascular plants that conducts fluid nutrients to the stems and roots.

phospholipid
a lipid containing phosphorus, the basic constituent of cell membranes.

phosphorylation
the addition of phosphate to an organic compound.

physiology
the study of the normal processes of living organisms, rather than their anatomical structure or biochemical composition.

plasmodium
a protoplasmic mass containing several nuclei and surrounded by a membrane.

polymerase
a term for any enzyme that catalyzes the lengthening of a polymeric molecule, such as nucleotides to polynucleotides.

predation
the act of one animal obtaining food by hunting another animal.

prostaglandin
any of a class of organic compounds present in many tissues, and derived from essential fatty acids. They have a variety of effects including vasodilation, vasoconstriction, and stimulation of bronchial and intestinal smooth muscle.

protein
macromolecules made up of long sequences of amino acids.

proximal
nearest the center or point of origin.

pyruvate
an organic acid produced during glycolosis and essential to the Kreb's cycle.

rain shadow
dry region in the protective shadow of tall mountain ranges.

respiration
in a cell, a chemical reaction that breaks down molecules from food to produce energy.

ribonucleic acid
A macromolecule in cells that plays an important role in protein synthesis. Commonly known as RNA.

saprotroph
an organism that obtains nourishment by absorbing dead organic material.

sarcomere
the functional unit of striated muscle, a segment of a myofibril between two membranes called Z lines.

sebum
the secretion of the sebaceous glands in the dermis.

sinus
a cavity or hollow space in bone or a saclike cavity in other tissue.

solute
the dissolved substance in a solution.

sphincter
a ring of muscle around an opening in a duct, or an orifice that closes the structure when constricted.

steroid
a group of lipids with a distinct four-ring nucleus.

sterol
a steroid-based alcohol.

subcutaneous
beneath the skin.

synapse
a junction between an axon of one nerve cell and the tip of a dendrite of another nerve cell. A nerve impulse is transmitted across the gap between them via a neurotransmitter.

tissue
a collection of similar cells that perform a particular function, such as muscle tissue.

twinning
the production of comparable structures by division.

vector
an organism, often a mosquito, fly, or tick that can carry and transmit an infectious agent to a vertebrate.

vesicle
a small sac usually containing a fluid substance.

xylem
tissue in vascular plants that conducts fluid upward from the roots and provides support for the plant.

zygote
a cell formed by the union of sperm and egg.

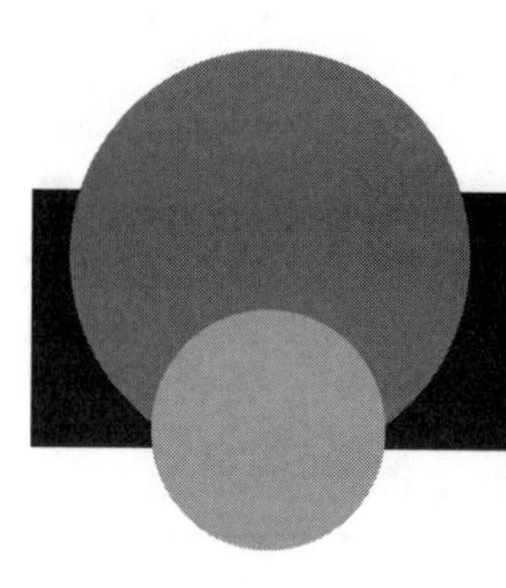

INDEX

Note: Page numbers in *italics* indicate figures.